LABORATORY BASICS

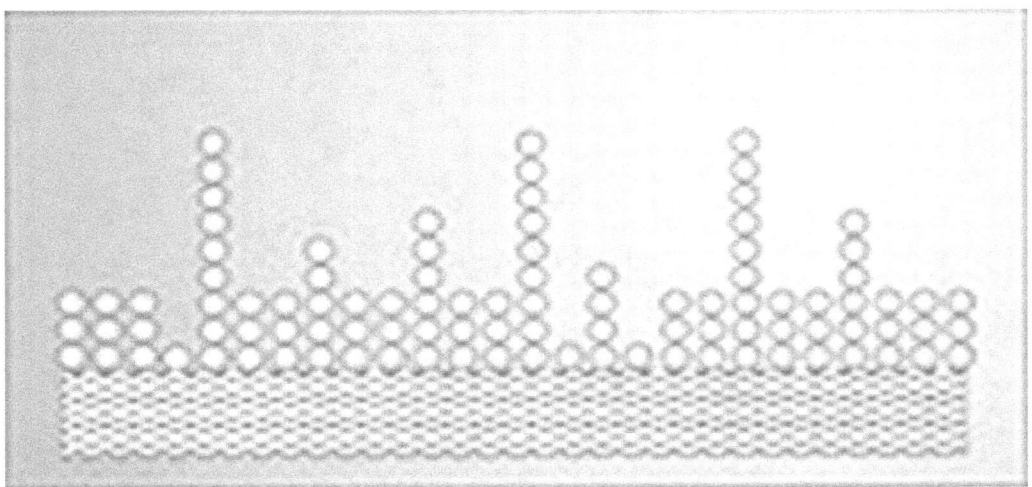

Dr. Mohammad Kabir Ahmed Associate Professor and Dr. Shakeruddin Ahmed, Senior Lecturer

University of Kuala Lumpur Royal College of Medicine Perak, 3, Green town, Ipoh, Perak, Malaysia

ISBN:1981903623
ISBN-13:9781981903627

i

DEDICATION

Dedicated to my family

PREFACE

Laboratory Basics differs from the usual text books or manual. But this is a guideline with some useful information regarding very basic areas of common technology. It is a foundation only. In coming future this book will be enriched by incorporating current standards and methods of analysis depending on the functionality of a laboratory. It is true that standard operating procedures (SOP) are classified control document of any laboratory. The standard operating procedure shall be written by the working laboratory personnel according to the quality manual and duly approved by a designated body of that laboratory. The SOPs incorporated here will serve as a guideline in terms of structure. It cannot be adopted by any laboratory as it is. These are not incorporated here with that intention. But basic structure of an SOP proposed here complies with the ISO 17025. This book is not just accumulation of disjointed unrelated topics.

I hope this book will become a basic source book in coming days for a laboratory having desire to get third party accreditation.

Universiti Kuala Lumpur
Royal College of Medicine Perak

Dr. Mohammad Kabir Ahmed

iii

CONTENTS

ACKNOWLEDGMENTS

An undertaking of this nature required co-operation and assistancenof many within and outside the department. I am thankful for their ungrudging help and inspiration to write this book on laboratory need. I would particularly like to acknowledge the precious support of my colleagues Dr. Syed Rahim bin Sayed, Dean, FOM, Dr. Noorzaid Muhammad and Dr.. Khyril Azwan bin Malim Zafaar for inspiration and needful official action.

Chemical Equilibrium

Different forms of particular chemical specie arises due to natural entropy but exist always in equilibrium. For example, pure water consists of the molecular compound (H_2O) and dissociated ions (H^2 and HO^-) that exist together in equilibrium:

$$H_2O_{(l)} \rightleftharpoons H^+_{(aq)} + OH^-_{(aq)}$$

The (l) subscript refers to the liquid state, and the (aq) subscript refers to ions in aqueous solution. Pure water at room temperature contains 1×10^{-7} M H^+ and OH^- .

The same is true for chemical reactions. The reactants will always exist in equilibrium with the products of the reaction.

At equilibrium the partial pressures or concentrations of the reactants and products are no longer changing, that is they have reached a steady state. It should be noted, that though the molecular level reactant molecules (or atoms or ions) are forming products, and product molecules are also returning to reactants. At equilibrium, the rate generation of products is equal to the reverse reaction of products going to reactants.

The Equilibrium Constant Expression

Equilibrium Constant

The equilibrium between reactants and products is described by an equilibrium constant. For the balanced reaction:

$$aA + bB \rightleftharpoons cC + dD$$

The equilibrium constant, K_{eq} is defined as:

$$K_{eq} = \frac{[C]^c [D]^d}{[A]^a [B]^b}$$

where the [ ] brackets indicate the concentration of the chemical species.
For the example of water,

$$H_2O_{(l)} \rightleftharpoons H^+_{(aq)} + OH^-_{(aq)}$$ the equilibrium constant is:

$$K_{eq} = \frac{[H^+] [OH^-]}{[H_2O]}$$

The concentration of water in a water solution is constant and this expression simplifies to:

$$K_w = (55.56 \text{ M}) * K_{eq} = [H^+][OH^-]$$

where K_w is called the dissociation constant of water and equals 1.00×10^{-14} at room temperature. The concentrations of $[H^+]$ and $[OH^-]$ therefore equal 1.00×10^{-7} M.

Acid/base equilibrium:

Most solutions containing different ions are in state of equilibrium - all concentrations are constant and not changing in time. This equilibrium is dynamic which means we have forward and reverse reactions going in the solution at the same rate, so that concentrations are not changing.

Let's assume acid HA, base BOH, and their analytical concentrations (ie concentrations off all forms present in solution) are respectively C_a and C_b.

For the acid dissociation reaction : $HA \leftrightarrow H^+ + A^-$ equilibrium is described by the **acid dissociation constant** defined as :

$$K_a = \frac{[H^+][A^-]}{[HA]} \quad 1.6$$

For the base dissociation reaction $BOH \leftrightarrow B^+ + OH^-$ **base dissociation constant** is defined as

$$K_b = \frac{[B^+][OH^-]}{[BOH]} \quad 1.7$$

Additional parameter used quite often to describe dissociation is a dissociation percentage (or dissociation fraction), defined as ratio of concentration of dissociated molecules to concentration of all molecules put into the solution. While not directly useful for finding equilibrium dissociation percentage is often used to validate assumptions done during calculations.

$$f_p = \frac{[A^-]}{C_a} \quad 1.8$$

It is commonly believed that strong acids and strong bases are fully dissociated.

Finally we will often need water ionization constant:

$$K_w = [H^+][OH^-] \quad 1.9$$

pH and pOH defined in the pH definition section can be used not only as measure of the ions concentration, but also in calculations. Let's take logarithm of both sides of 1.4 equation:

$$\log(K_w) = \log([H^+]) + \log([OH^-])_{1.10}$$

Change signs and use p notation: $pK_w = pH + pOH_{1.6}$

Equation 1.6 is often used when we need to calculate pH but it is much easier to calculate pOH and vice versa.

$$K = \frac{[H^+][OH^-]}{[H_2O]} \quad_{1.11}$$

Brønsted-Lowry's acids and bases

As all reactions we are interested in take place in and water dissociates itself into H^+ and OH^- ions, classic definition of acid as a substance that dissociates producing H^+ ions. Consider the solution of salt of weak base BOH. Such solution contains B^+ ions that are between products of BOH dissociation:

$$BOH \leftrightarrow B^+ + OH^-$$

with equilibrium described by the already mentioned in the previous section base dissociation constant:

$$K_b = \frac{[B^+][OH^-]}{[BOH]} \quad_{2.1}$$

For the equilibrium BOH molecules are needed. As there are already OH^- ions from the water dissociation present in the solution, they will react with B^+. This will lower OH^- concentration, forcing water to dissociate further. Final solution in equilibrium will contain some BOH molecules and some excess of H^+ - so it will be acidic, even if we haven't add any acid.

According to Brønsted and Lowry, acid is a substance that can donate the proton and base is a substance than can accept the proton. The most important outcome of this definition is the fact that every acid loosing its proton becomes a Brønsted-Lowry base and that every base when protonated becomes a Brønsted-Lowry acid. These pairs of acid and base are called conjugate. In other words every acid loosing proton becomes its conjugate base, and every protonated base becomes its conjugate acid.

This approach has some interesting implications. Let's take a reaction of conjugate base A^- with water:

$$A^- + H_2O \rightarrow HA + OH^-$$

Its equilibrium constant is

$$K_b = \frac{[HA][OH^-]}{[A^-]} \quad 2.2$$

Multiplying this equation by the equation for acid dissociation constant we get

$$K_a K_b = \frac{[H^+][A^-]}{[HA]} \frac{[HA][OH^-]}{[A^-]} \quad 2.3$$

[A⁻] and [HA] cancel out leaving $\quad K_a K_b = [H^+][OH^-] = K_w \, 2.4$

Or $\quad pK_a + pK_b = pK_w \, 2.5$

Acid-base indicator:
An acid-base indicator is a weak acid or a weak base. Indicators have a very useful property - they change color depending on the pH of the solution they are in. This color change is not at a fixed pH, but rather, it occurs gradually over a range of pH values. This range is termed the color change interval. Each pH indicator is defined by a useful pH range. For example Phenolphthalein changes from colorless at 8.0 to pink at 10.0. And Bromthymol Blue has a useful range from 6.0 (yellow) to 7.6 (blue).

Preparing Chemical Solutions
Lab experiments and types of research often require preparation of chemical solutions in their procedure. We look at preparation of these chemical solutions by weight (w/v) and by volume (v/v). The glossary below cites definitions to know when your work calls for making these and the most accurate molar solutions.

To this we add information designed for understanding how to use the pH scale when measuring acidity or alkalinity of a solution.

Titration:
Titration is the quantitative measurement of an analyte in solution by completely reacting it with a reagent.
The point at which all of the analyte is consumed is called the endpoint and is determined by some type of indicator that is also present in the solution. For acid-base titrations, indicators are available that change color when the pH changes. When all of the analyte is neutralized, further addition of the titrant

causes the pH of the solution to change causing the color of the indicator to change.

The analyte concentration is calculated from the reaction stoichiometry and the amount of reagent that was required to react with all of the analyte.

Instrumentation

Manual titration is done with a burette, which is a long graduated tube to hold the titrant. The amount of titrant used in the titration is found by reading the volume of titrant in the burette before beginning the titration and when the endpoint is reached, and taking the difference. The most important factor for making accurate titrations is to read the burette volumes reproducibly. The figure shows how to do so by using the bottom of the meniscus to read the reagent volume in the burette.

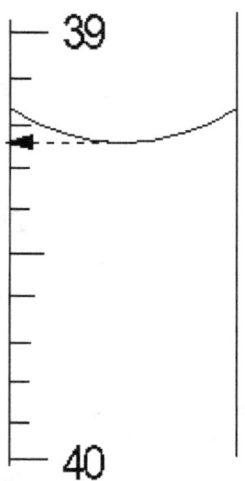

For repetitive titrations, autotitrators with microprocessors are available that deliver the titrant, stop at the endpoint, and calculate the concentration of the analyte. The endpoint is usually detected by some type of electrochemical measurement. Some examples of titrations for which autotitrators are available include:

Electrochemistry:

Electrochemistry can be broadly defined as the study of charge-transfer phenomena. As such, the field of electrochemistry includes a wide range of different chemical and physical phenomena. These areas include: battery chemistry, photosynthesis, ion-selective electrodes, coulometry, and many biochemical processes. Although wide ranging, electrochemistry have many practical applications in analytical measurements. Presently this phenomena is being in next generation sequencing procedure of DNA.

Electroanalytical chemistry

A good working definition of the field of electroanalytical chemistry would be that it is the field of electrochemistry that utilizes the relationship between chemical phenomena which involve charge transfer (e.g. redox reactions, ion separation, etc.) and the electrical properties that accompany these phenomena for some analytical determination.

Potentiometry involves the measurement of potential for quantitative analysis, and electrolytic electrochemical phenomena involve the application of a potential or current to drive a chemical phenomenon, resulting in some measurable signal which may be used in an analytical determination.

Potentiometry

Potentiometry is the field of electroanalytical chemistry in which potential is measured under the conditions of no current flow. The measured potential may then be used to determine the analytical quantity of interest, generally the concentration of some component of the analyte solution. The potential that develops in the electrochemical cell is the result of the free energy change that would occur if the chemical phenomena were to proceed until the equilibrium condition has been satisfied.

2. INTERNATIONAL SYSTEM UNITS AND SOLUTIONS

International System of units (SI): Prefixes used in Metric system

Factor	Prefix	Symbol	Factor	Prefix	Symbol
10^{-18}	atto	a	10	deca	da
10^{-15}	femto	f	10^2	hecto	h
10^{-12}	pico	p	10^3	kilo	k
10^{-9}	nano	n	10^6	mega	M
10^{-6}	micro	μ	10^9	giga	G
10^{-3}	milli	m	10^{12}	tera	T
10^{-2}	centi	c			
10^{-1}	deci	d			

Units of Mass and Volume:

Unit of mass is the kilogram (kg). The gram (g) is defined as $1/1000$ (10^{-3}) of the mass international prototype kilogram.

Unit of volume is the liter (l). Here liter is defined as one cubic decimeter.

Mass			Volume		
Name	Symbol	Multiples & Fraction of a gm	Name	Symbol	Multiples & Fractions of one liter
killogram	kg	1000	kiloliter	kl	1000
hectogram	hg	100	hectoliter	hl	100
decagram	dkg (dag)	10	decaliter	dkl (dag)	10
gram	g	1	liter	l	1
decigram	dg	0.1	deciliter	dl	0.1
centigram	cg	0.01	centiliter	cl	0.01
milligram	mg	0.001	millilioter	ml	0.001
micrigram	μg	0.000001	microliter	μl	0.000001

Solutions:

Solute - The substance which dissolves in a solution
Solvent - The substance which dissolves another to form a solution. For example, in a sugar and water solution, water is the solvent; sugar is the solute.
Solution - A mixture of two or more pure substances. In a solution one pure substance is dissolved in another pure substance homogenously. For example, in a sugar and water solution, the solution has the same concentration throughout, i.e. it is homogenous.

Mole - A fundamental unit of mass (like a "dozen" to a baker) used by chemists. This term refers to a large number of elementary particles (atoms, molecules, ions, electrons, etc) of any substance. 1 mole is 6.02×10^{23} molecules of that substance. (Avogadro's number). M

Preparation of solutions:

Many experiments involving chemicals call for their use in solution form. That is, two or more substances are mixed together in known quantities. This may involve weighing a precise amount of dry material or measuring a precise amount of liquid. Preparing solutions accurately will improve an experiment's safety and chances for success.

Solution 1: Using percentage by weight (w/v)

Formula: The formula for weight percent (w/v) is: [Mass of solute (g) / Volume of solution (ml)] x 100

Example: A 10% NaCl solution has ten grams of sodium chloride dissolved in 100 ml of solution.

Procedure: Weigh **10g** of sodium chloride. Pour it into a graduated cylinder or volumetric flask containing about **80ml** of water. Once the sodium chloride has dissolved completely (swirl the flask gently if necessary), add water to bring the volume up to the final 100 ml. Caution: Do not simply measure **100ml** of water and add 10g of sodium chloride. This will introduce error because adding the solid will change the final volume of the solution and throw off the final percentage.

Solution 2: Using percentage by volume (v/v)

When the solute is a liquid, it is sometimes convenient to express the solution concentration as a volume percent.

Formula

The formula for volume percent (v/v) is: [Volume of solute (ml) / Volume of solution (ml)] x 100

Example

Make 1000ml of a 5% by volume solution of ethylene glycol in water.

Procedure

First, express the percent of solute as a decimal: 5% = 0.05
Multiply this decimal by the total volume: 0.05 x 1000ml = 50ml (ethylene glycol needed).
Subtract the volume of solute (ethylene glycol) from the total solution volume:

1000ml (total solution volume) - 50ml (ethylene glycol volume) = 950ml (water needed)

Dissolve **50ml** ethylene glycol in a little less than **950ml** of water. Now bring final volume of solution up to **1000ml** with the addition of more water. (This eliminates any error because the final volume of the solution may not equal the calculated sum of the individual components).

So, 50ml ethylene glycol / 1000ml solution x100 = 5% (v/v) ethylene glycol solution.

Solution 3: Molar Solutions

Molar solutions are the most useful in chemical reaction calculations because they directly relate the moles of solute to the volume of solution.

Formula

The formula for molarity (M) is: moles of solute / 1 liter of solution or gram-molecular masses of solute / 1 liter of solution.

Examples

The molecular weight of a sodium chloride molecule (NaCl) is 58.44, so one gram-molecular mass (=1 mole) is 58.44 g. From the atomic mass (or weight) of Na is 22.99, the atomic mass of Cl is 35.45,

therefore NaCl is (22.99 + 35.45) = 58.44.

58.44g of NaCl dissolved in a final volume of **1 liter**, will make a **1M NaCl** solution or a 1 molar solution.

Procedure: To make molar NaCl solutions of other concentrations dilute the mass of salt to 1000ml of solution as follows:

0.1M NaCl solution requires **0.1 x 58.44 g of NaCl = 5.844g**

0.5M NaCl solution requires **0.5 x 58.44 g of NaCl = 29.22g**

2M NaCl solution requires **2.0 x 58.44 g of NaCl = 116.88g**

3 PH METER

pH Meter:

The pH meter measures the pH of a solution using an ion-selective electrode (ISE) that responds to the H^+ concentration of the solution. The pH electrode produces a voltage that is proportional to the concentration of the H^+ concentration, and making measurements with a pH meter is therefore a form of potentiometry. The pH electrode is attached to control electronics which convert the voltage to a pH reading and displays it on a meter.

Instrumentation:

A pH meter consists of a H^+-selective membrane, an internal reference electrode, an external reference electrode, and a meter with control electronics and display. Commercial pH electrodes usually combine all electrodes into one unit that are then attached to the pH meter.

Picture of a pH meter

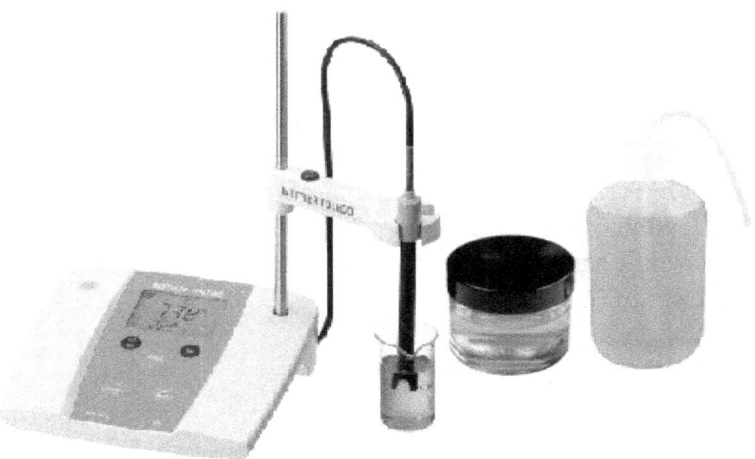

Electrolytic Methods

Unlike potentiometry, where the free energy contained within the system generates the analytical signal, electrolytic methods are an area of electroanalytical chemistry in which an external source of energy is supplied to drive an electrochemical reaction which would not normally occur. The externally applied driving force is either an applied potential or current. When potential is applied, the resultant current is the analytical signal; and when current is applied, the resultant potential is the analytical signal. Techniques which utilize applied potential are typically referred to as voltammetric methods while those with applied current are referred to as galvanostatic methods.

Voltammetry Methods

Voltammetry refers to the measurement of current that results from the application of potential. Unlike potentiometric measurements, which employ only two electrodes, voltammetric measurements utilize a three electrode electrochemical cell. The use of the three electrodes (working, auxiliary, and reference) along with the potentiostat instrument allows accurate application of potential functions and the measurement of the resultant current. The different voltammetric techniques that are used are distinguished from each other primarily by the potential function that is applied to the working electrode to drive the reaction, and by the material used as the working electrode.

Toward Understanding pH

Let's begin with the words acidic and basic as extremes which describe solutions as hot and cold are extremes which describe temperature. Just as mixing hot and cold water evens out the temperature, mixing acids and bases can cancel their extreme effects and is then considered neutral.

The pH scale can tell if a liquid is more acid or more base, just as the Fahrenheit or Celsius scale is used to measure temperature. The range of the pH scale is from 0 to 14 from very acidic to very basic. A pH of 7 is neutral. A pH less than 7 is acidic and greater than 7 is basic.

Each whole pH value below 7 is ten times more acidic than the next higher value. For example, a pH of 4 is ten times more acidic than a pH of 5 and a hundred times (10 X 10) more acidic than a pH of 6. This holds true for pH values above 7, each of which is ten times more basic (also called alkaline) than the next lower whole value. An example would be, a pH of 10 is ten times more alkaline than a pH of 9.

Acid-Base Indicators

Liquid pH indicators are used to test other solutions. A few drops of the right indicator added to an unknown solution can tell you its pH value. Chemists use pH indicators in a common laboratory procedure called titration. Here, an unknown substance is measured by carefully adding a solution of known concentration until a neutral point is reached. The neutral point is indicated by the color change of a pH indicator mixed in with the unknown solution.

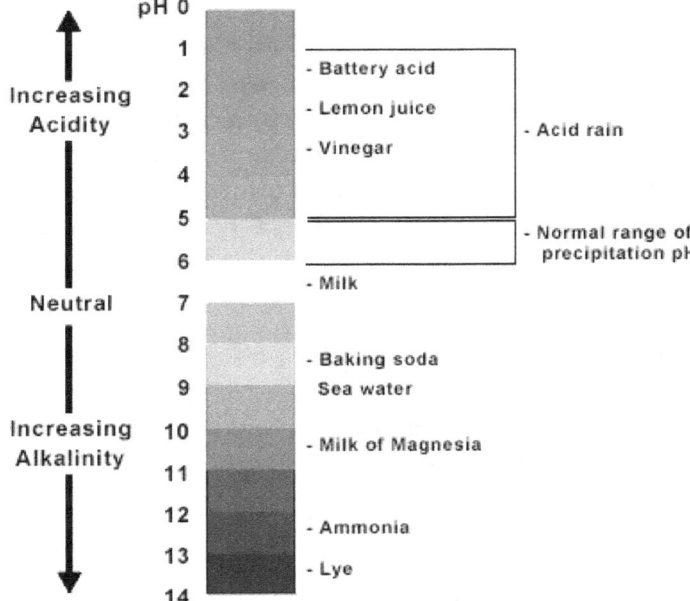

The pH Scale
With digital readout via electronic meter:
pH is measured in various fields, that includes environmental, agricultural, wastewater, pharmaceutical, and educational applications.
Electronic, benchtop meters are available that read pH to resolutions of 0.001. Portable, pH meters are suitable for a wide array of testing needs.with accuracies of 0.1 or 0.01 pH.Typical standards solutions of pH 4.01, pH 7.01 and pH 10.01, are used for calibration of pH meter..

pH definition, pOH definition

Every water solution contains H^+ ions. Their concentration is one of the most important parameters describing solution properties.
Concentrations of H^+ can change in a very wide range, it can be 10 M as well as 10^{-15} M. Such numbers are inconvenient to use so to simplify things Danish biochemist Søren Sørensen developed in 1909 the pH scale and introduced **pH definition** - minus logarithm base 10 of $[H^+]$:

$$pH = -\log([H^+])_{1.4}$$

It is much easier to use pH definition and to say "pH of the solution is 4.1" than to use concentrations - as in "H^+ concentration is 0.000079M".

Not only H^+ ions are present in every water solution. Also OH^- ions are always present, and their concentration can change in the same very wide range. Thus it is also convenient to use similar definition to describe [OH-].

$$pOH = -\log([OH^-])_{1.5}$$

pH scale

Acidity of the solution is so important, that it was convenient to create a special **pH scale** for its measurements. This pH scale uses Sørensen's pH definition. Concentration of H^+ is usually confined to $1\text{-}10^{-14}M$ range. Thus pH scale contains values falling between 0 and 14. In some rare cases you may see pH lower than 0 or higher than 14, when the concentration of H^+ take some extreme values.

On the pH scale pure water has pH 7, although you will probably never see water pure enough for such pH. Air always contains small amounts of carbon dioxide which dissolves in water making it slightly acidic - with pH of about 5.7.
pH scale is used very extensively in the chemistry, biochemistry and biology.

	Acidic							Neutral
H^+	1.10^0	1.10^1	1.10^2	1.10^3	1.10^4	1.10^5	1.10^6	1.10^7
pH	0	1	2	3	4	5	6	7
pOH	14	13	12	11	10	9	8	7
OH^-	1.10^{14}	1.10^{13}	1.10^{12}	1.10^{11}	1.10^{10}	1.10^9	1.10^8	1.10^7

	Alkaline							
H^+	1.10^8	1.10^9	1.10^{10}	1.10^{11}	1.10^{12}	1.10^{12}	1.10^{14}	
pH	8	9	10	11	12	13	14	
pOH	6	5	4	3	2	1	0	
OH^-	1.10^6	1.10^5	1.10^4	1.10^3	1.10^2	1.10^1	1.10^0	

There are two things worth of remembering about pH scale. First, as pH scale is logarithmic 1 unit pH change means tenfold change in the H^+ ion concentration. Second, while only solution with pH=7.00 is strictly neutral, all solutions with pH in the range 4-10 have concentration of H^+ and OH^- lower than $10^{-4}M$ - which can be easily influenced with small additions of acid and base.

pH electrode: pH meter is a precise voltameter connected to the ion selective electrode. Voltage produced by the pH electrode is proportional to logartithm of the H+ activity. pH meter voltameter display is scaled in such a way that the displayed result of measurement is just the pH of the solution.

Commercial pH electrodes used in pH meters consist of a H+ selective membrane or a very thin glass, and internal and external reference electrodes, usually combined in one housing.

substance	pH scale value
battery acid	pH below 1
gastric juice	pH about 2
orange juice	pH between 3 and 4
milk	pH about 6.5
blood	pH between 7.34 and 7.45, very precisely kept in this range, as the correct pH is crucial for the survival
normal soap	pH about 9-10
bleach	pH about 12.5

Using properly calibrated pH meter with a good electrode one may measure pH with +/- 0.01 unit accuracy without any problem. Some typical pH Values:

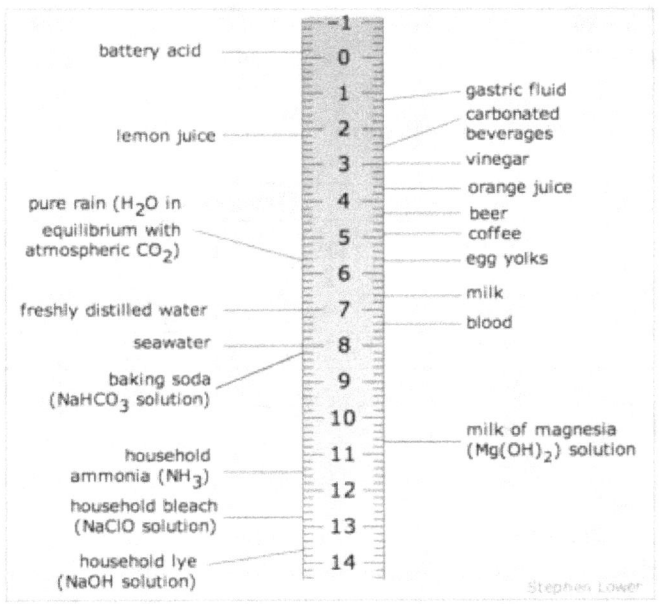

Standard buffers for calibration of pH meter:

Temp. (^{0}C)	0.1M HCl	Saturated Potassium Hydrogen tartrate	0.05M Potassium hydrogen phthalate	0.05M phosphate	0.01 Borax	Aturated calcium hydroxide
0	1.10		4.003.	6.984	9.464	13.423
5	1.10		3.999	6.951	9.395	13.207
10	1.10		3.998	6.923	9.332	13.003
15	1.10		3.999	6.900	9.276	12.810
20	1.10		4.002	6.881	9.225	12.627
25	1.10	3.557	4.008	6.865	9.180	12.454
30	1.10	3.552	4.0156	6.853	9.139	12.289
35	1.10	3.549	4.024	6.844	9.102	12.133
38	1.10	3.548	4.030	6.840	9.081	12.043
40	1.10	3.547	4.035	6.838	9.068	11.984
45	1.10	3.547	4.047	6.834	9.038	11.841
50	1.10	3.549	4.060	6.833	9.011	11.705
55	1.11	3.554	4.075	6.834	8.985	11.574
60	1.11	3.560	4.091	6.836	8.962	11.449
70	1.11	3.580	4.126	6.845	8.921	
80	1.11	3.609	4.164	6.859	8.885	
90	1.12	3.650	4.205	6.877	8.850	
95	1.12	3.674	4.227	6.886	8.833	

pH values of standard solutions:

Normality	pH Values			
	HCl	CH_3COOH	NaOH	NH_3
1	0.10	2.37	14.05	11.77
0.1	1.07	2.87	13.07	11.27
0.01	2.02	3.37	12.12	10.77
0.001	3.01	3.87	11.13	10.27
0.0001	4.01			

Universal Indicators:

An approximate measure of pH may be obtained by using a pH indicator. A pH indicator is a substance that changes colour around a particular pH value. It is a weak acid or weak base and the colour change occurs around 1 pH unit either side of its acid dissociation constant, or pK_a, value. For example, the naturally occurring indicator litmus is red in acidic solutions (pH<7) and blue in alkaline (pH>7) solutions. Universal indicator consists of a mixture of indicators such that there is a continuous colour change from about pH 2 to pH 10. Universal indicator paper is simple paper that has been impregnated with universal indicator.

Universal indicator components

Indicator	Low pH color	Transition pH range	High pH color
Thymol blue (first transition)	red	1.2–2.8	orange
Methyl red	red	4.4–6.2	yellow
Bromothymol blue	yellow	6.0–7.6	blue
Thymol blue (second transition)	yellow	8.0–9.6	blue
Phenolphthalein	colorless	8.3–10.0	purple

A solution whose pH is 7 is said to be neutral, that is, it is neither acidic nor basic. Water is subject to a self-ionisation process.

$$H_2O \rightleftharpoons H^+ + OH^-$$

4 OPTICS TO SPECTROSCOPY

Optics & Optical Materials

Successfully applying spectroscopy to scientific problems requires a basic understanding of the components that are available to produce and control electromagnetic radiation. This outline leads to documents that discuss basic optics, light and excitation sources, wavelength separators, and light detectors.

Mirrors
X-rays
Ultraviolet   aluminum
Visible   aluminum
Near infrared gold
infrared   copper, gold

Lenses
X-rays
Ultraviolet   fused silica (synthetic quartz)
Visible   glass
Near infrared glass
infrared   ZnSe

Windows
X-rays   beryllium
Ultraviolet   fused silica (synthetic quartz)
Visible   glass
Near infrared glass
infrared   ZnSe, NaCl, BaF_2

Mirrors
Flat Mirrors

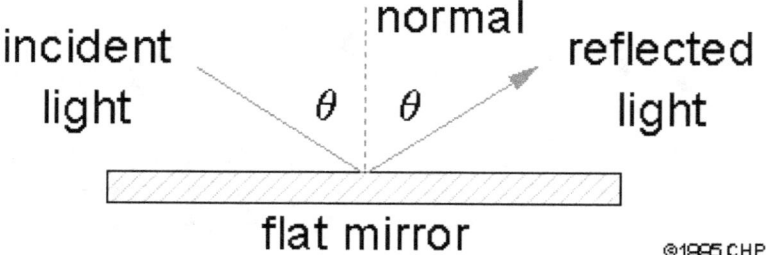

Spherical Mirrors
Spherical mirror: for radius of curvature, R, the focal length, f, is R/2.
Light collection f-number: $f/\# = l/d$ where l is distance and d is diameter of the mirror.

Refraction
When light enters a solid material from air, the speed at which the light travels decreases. If the light enters at an off-normal angle, the direction of the light changes. The light is said to be refracted and the refracted angle is given by Snell's law: $n_1 \sin \theta_1 = n_2 \sin \theta_2$

Reflection and refraction of an incident light ray at a surface

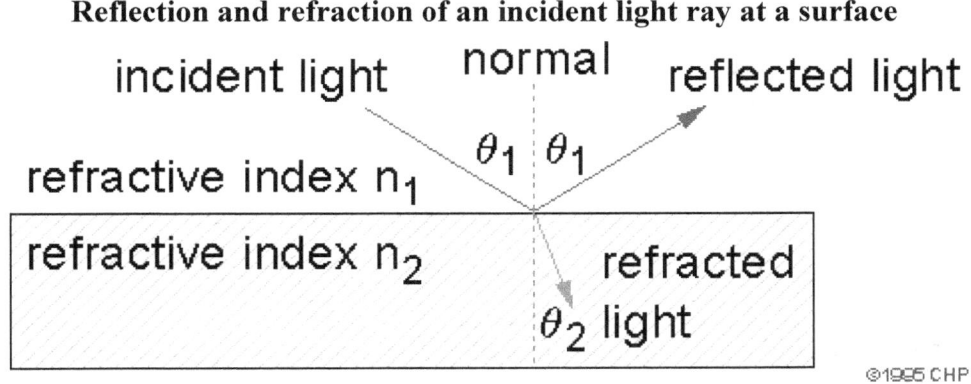

Prisms
A prism is a transparent optic that is shaped to bend light. Since the refractive index of a material varies with wavelength, prisms are useful for dispersing different wavelengths of light.

Dispersion of white light by a prism:

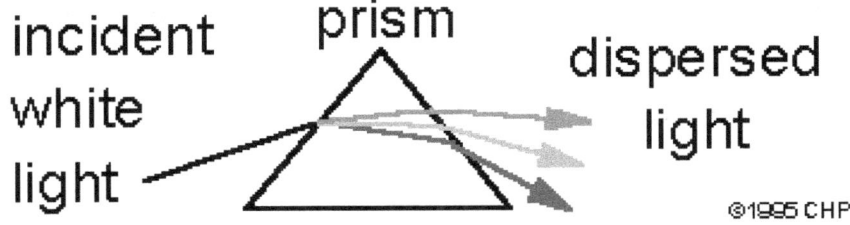

Lenses
Lenses are transparent optical components that use refraction to focus or collimate electromagnetic radiation.

Properties of lenses

Lenses are characterized by their focal length. The focal length is given by the lens maker's formula: $1/f(lambda) = [n(lambda) - 1][1/R_1 - 1/R_2]$
where f is the focal length, n is refractive index, and R_1 and R_2 are the radii of curvature of the two surfaces of the lens.

Illustration of lens properties:

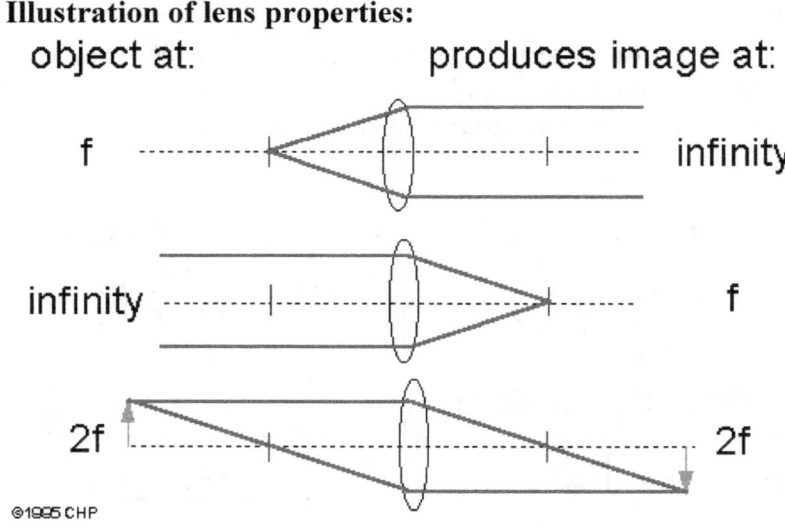

Focus spot size of light from infinity: spot diameter = lambda * f / pi * D where lambda is wavelength, f is focal length, and D is diameter of the incoming beam.

Images and ray tracing

The position of an image can be determined by tracing three lines in a diagram: parallel to the optical axis and through the focal point through the center of the lens through the focal point to the lens and then parallel to the optical axis.

Finding an image by ray tracing:

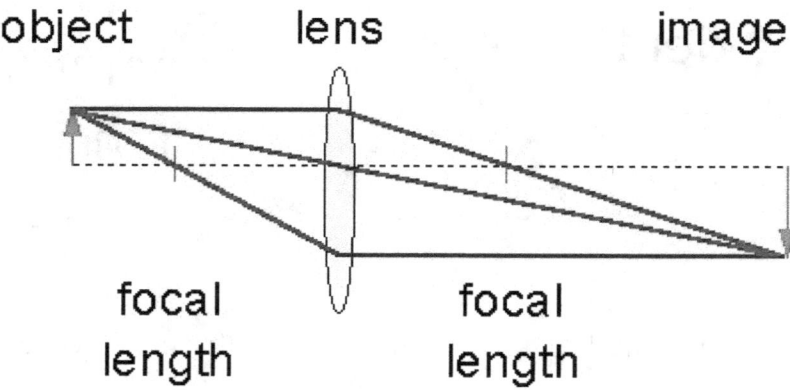

The position of an image can be found from the ray trace or from: $1/f = 1/x_o + 1/x_i$ where f is the lens focal length, x_o is distance of the object from the lens, and x_i is the distance of the image from the lens.

Magnification, M, is the ratio of the image size to the object size, and is equal to: $M = x_i / x_o$ where x_i and x_o are the distances of the image and object from the lens, respectively.

Light collection

The light collection efficiency is the solid angle that an optic makes with an object. The f-number describes this angle: f-number: $f/\# = l/d$ where l is distance and d is diameter of lens.

The solid angle that a lens collects is approximately: $OMEGA = pi\ d^2 / 4\ l^2$

The fraction of light that an optic collects is this solid angle divided by the total 4 piste radians: fraction collected = $OMEGA /4pi = pi\ d^2 / 16\ pi\ l^2 = 1 / 16\ (f/\#)^2$.

Aberrations

Lenses (and curved mirrors) do not focus light perfectly. Chromatic and spherical aberations occur on-axis and coma and astigmitism occur off-axis.

Chromatic aberration

Chromatic aberration occurs due to the variation of refractive index with wavelength for a lens material (there is no chromatic aberation in curved mirrors). This wavelength dependence results in slightly different focal lengths for different wavelengths of light. Compound lenses, called achromats, can reduce or eliminate chromatic aberation because the components are chosen such that the variation in refractive index as a function of wavelength cancels out.

Spherical aberration

Spherical aberation results because the actual focal point of a light ray depends on its distance from the optic axis.

Coma

Coma is caused by the distortion of a wavefront as it encounters an optic asymmetrically. The result for collimated incoming light is a circle instead of a point image. The light rays farther from the optic axis have more severe aberation and the resulting image looks like a comet-shaped series of circles.

Astigmatism

The projection of an optic off-axis looks squashed in one direction. The squashed direction focuses light to a greater extent than the normal dimension. The result is two line images.

Minimizing aberrations

work on or near the optic axis use compound lenses (achromats, doublets, triplets) which can be designed to reduce chromatic aberation, spherical aberation, and coma use computer optimized aspheric lenses

Polarizers

Unpolarized electromagnetic radiation has electric field vectors in all directions perpendicular to the direction of light propagation. Polarizers isolate one component of the electric field vector by selectively transmitting or reflecting one specific electric field direction.

Electromagnetic Radiation

Electromagnetic radiation is an energy wave that is composed of an electric field component and a magnetic field component. The electric and magnetic fields are orthogonal to each other and orthogonal to the direction of propogation of the wave.

Type of Spectrum and their interaction with matter is given below:

The Visible Spectrum

infrared
light

ultraviolet
light

| | | | |
700   600   500   400

Wavelength (nm)

The wavelength is the length of one complete oscillation and the frequency is the number of oscillations per second. Electromagnetic waves travel through a vacuum at 2.99792×10^8 m/s. The relation between speed of light (c), wavelength (*lambda*), and frequency (*nu*) is: c = *lambda * nu*

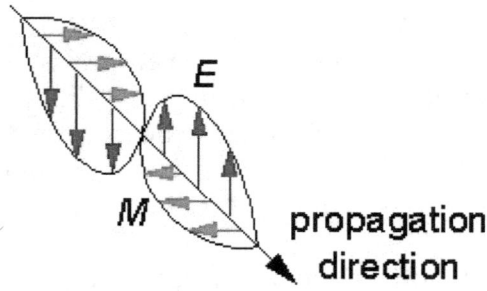

Range of the Spectrum- Electromagnetic Waves:

Region of the Spectrum	Main interactions with matter
Radio	Collective oscillation of charge carriers in bulk material (plasma oscillation). An example would be the oscillation of the electrons in an antenna.
Microwave through far infrared	Plasma oscillation, molecular rotation
Near infrared	Molecular vibration, plasma oscillation (in metals only)
Visible	Molecular electron excitation (including pigment molecules found in the human retina), plasma oscillations (in metals only)
Ultraviolet	Excitation of molecular and atomic valence electrons, including ejection of the electrons (photoelectric effect)
X-rays	Excitation and ejection of core atomic electrons
Gamma rays	Energetic ejection of core electrons in heavy elements, excitation of atomic nuclei, including dissociation of nuclei
High energy gam ma rays	Creation of particle-antiparticle pairs. At very high energies a single photon can create a shower of high energy particles and antiparticles upon interaction with matter.

Wave-particle duality

Electromagnetic radiation shows both wave and particle characteristics depending on how the radiation is observed. Einstein first postulated that the energy of radiation is quantized and that radiation is composed of energy packets that were later named photons. The energy (E) of a photon depends on its frequency (or wavelength): $E = h * nu = h * c / lambda$ where h is Planck's constant (6.62618×10^{-34} Js), nu is the frequency of the radiation (Hz), c is the speed of light (2.99792×10^{8} m/s), and $lambda$ is wavelength (m).

de Broglie equation

Analogous to radiation, particles; such as electrons, protons, and neutrons have wave properties as determined by the de Broglie equation: *lambda* = h/p where *lambda* is wavelength, h is Planck's constant, and p is the momentum of the particle.

Electromagnetic Spectrum

Type of Radiation	Frequency Range (Hz)	Wavelength Range	Type of Transition
gamma-rays	10^{20}-10^{24}	<10^{-12} m	nuclear
x-rays	10^{17}-10^{20}	1 nm-1 pm	inner electron
ultraviolet	10^{15}-10^{17}	400 nm-1 nm	outer electron
visible	4-7.5x10^{14}	750 nm-400 nm	outer electron
near-infrared	10^{12}-4x10^{14}	2.5 um-750 nm	outer electron molecular vibrations
infrared	10^{11}-10^{12}	25 um-2.5 um	molecular vibrations
microwaves	10^{8}-10^{12}	1 mm-25 um	molecular rotations, electron spin flips*
radio waves	10^{0}-10^{8}	>1 mm	nuclear spin flips*

*energy levels split by a magnetic field

Photomultiplier Tube (PMT)

Photomultiplier Tubes (PMTS) are light detectors that are useful in low intensity applications such as fluorescence spectroscopy. Due to high internal gain, PMTs are very sensitive detectors.

Design

PMTs are similar to phototubes. They consist of a photocathode and a series of dynodes in an evacuated glass enclosure. Photons that strikes the photoemissive cathode emits electrons due to the photoelectric effect. Instead of collecting these few electrons at an anode like in the phototubes, the electrons are accelerated towards a series of additional electrodes called dynodes. These electrodes are each maintained at a more positive potential. Additional electrons are generated at each dynode. This cascading effect creates 10^5 to 10^7 electrons for each photon hitting the first cathode depending on the number of dynodes and the accelerating voltage. This amplified signal is finally collected at the anode where it can be measured.

Typical specifications
Wavelength range: 110-1100 nm (wavelength sensitivity dependent on wavelength, uv-sensitive PMTs must have uv-transmitting windows.)
Quantum efficiency (Q.E., number of electrons ejected by the photocathode / number of incident photons): 1-10%
Response time: 1-15 ns

Typical specifications
Wavelength range: 110-1100 nm (wavelength sensitivity dependent on wavelength, uv-sensitive PMTs must have uv-transmitting windows, see optical materials)
Quantum efficiency (Q.E., number of electrons ejected by the photocathode / number of incident photons): 1-10%
Response time: 1-15 ns

Schematic of a PMT

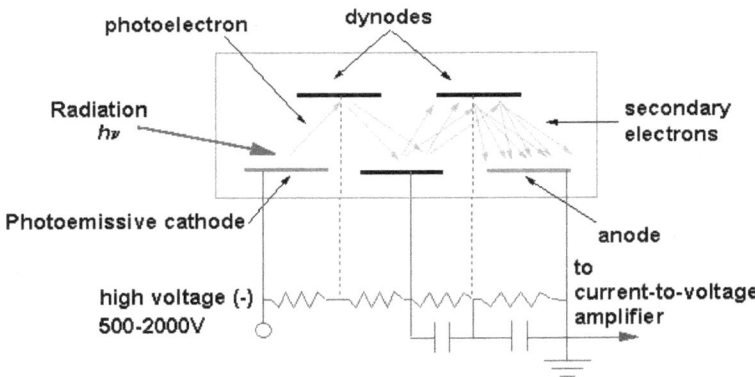

Lamps
Lamps convert electrical energy into radiation. Different designs and materials are needed to produce light in different parts of the electromagnetic spectrum. The following sections describe several different types of lamps that are useful in spectroscopy.

Blackbody Sources
A hot material, such as an electrically-heated filament in a light bulb, emits a continuum spectrum of light. The spectrum is approximated by Planck's radiation law for blackbody radiators:
$$B = \{2hnu^3/c^2\}\{1/\exp(hnu/kT)-1\}$$
The most common incandescent lamps and their wavelength ranges are: tungsten filament .
Lamps : 350 nm - 2.5 um

Glowbar : 1 - 40 um
Nernst glower : 400 nm - 20 um
Tungsten lamps are used in visible and near-infrared (NIR) absorption spectroscopy and the glowbar and Nernst glower are used for infrared spectroscopy.

Discharge Lamps

Discharge lamps, such as neon signs, pass an electric current through a rare gas or metal vapor to produce light. The electrons collide with gas atoms, exciting them to higher energy levels which then decay to lower levels by emitting light. Low-pressure lamps have sharp line emission characteristic of the atoms in the lamp, and high-pressure lamps have broadened lines superimposed on a continuum. Common discharge lamps and their wavelength ranges are: hydrogen or deuterium : 160 - 360 nm, mercury : 253.7 nm, and weaker lines in the near-uv and visible Ne, Ar, Kr, Xe discharge lamps : many sharp lines throughout the near-uv to near-IR, xenon arc : 300 - 1300 nm
Deuterium lamps are the uv source in uv-vis absorption spectrophotometers. The sharp lines of the mercury and rare gas discharge lamps are useful for wavelength calibration of optical instrumentation. Mercury and xenon arc lamps are used to excite fluorescence.

Hollow-cathode Lamps

Hollow-cathode lamps are a type of discharge lamp that produce narrow emission from atomic species. They get their name from the cup-shaped cathode, which is made from the element of interest. The electric discharge ionizes rare gas atoms, which are accelerated into the cathode and sputter metal atoms into the gas phase. Collisions with gas atoms or electrons excite the metal atoms to higher energy levels, which decay to lower levels by emitting light.

Schematic of a hollow-cathode lamp

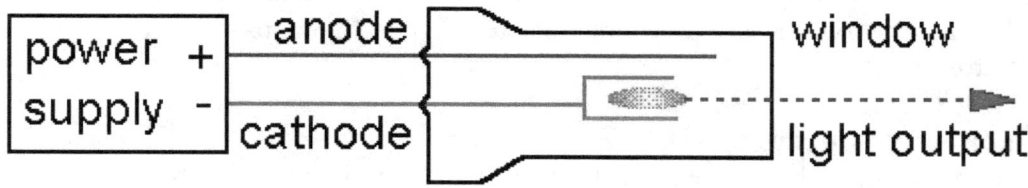

Hollow-cathode lamps have become the most common light source for atomic absorption (AA) spectroscopy. They are also sometimes used as an excitation source for atomic-fluorescence spectroscopy (AFS).

Lasers

A laser is a coherent and highly directional radiation source. LASER stands for Light Amplification by Stimulated Emission of Radiation.

A laser consists of at least three components:

a gain medium that can amplify light that passes through it

an energy pump source to create a population inversion in the gain medium

two mirrors that form a resonator cavity

The gain medium can be solid, liquid, or gas and the pump source can be an electrical discharge, a flashlamp, or another laser. The specific components of a laser vary depending on the gain medium and whether the laser is operated continuously (cw) or pulsed. The following headings describe specific laser designs.

Gas Lasers

Gas lasers are typically excited by an electrical discharge.

excimer : ArF^* - 248 nm, $XeCl^*$ - 308 nm (pulsed) , nitrogen : 337 nm (pulsed)

He-Ne : 632.8 nm (cw) , Ar ion : 488, 541 nm (cw)

CO2 : 10.6 μm (cw or pulsed)

Schematic of a cw gas laser :

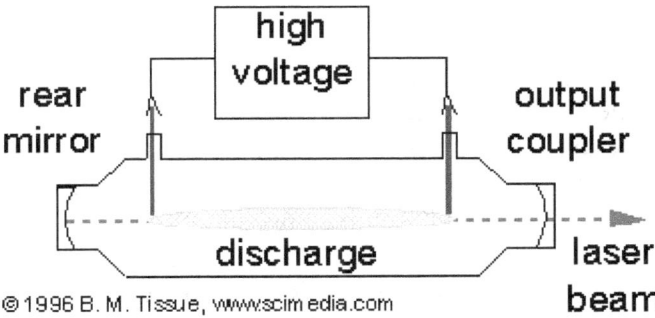

© 1996 B. M. Tissue, www.scimedia.com

Solid-State Lasers

The gain medium in a solid-state laser is an impurity center in a crystal or glass. Solid-state lasers made from semiconductors are described below. The first laser was a ruby crystal (Cr^{3+} in Al_2O_3) that lased at 694 nm when pumped by a flashlamp. The most commonly used solid-state laser is one with Nd^{3+} in a $Y_3Al_5O_8$ (YAG) or $YLiF_4$(YLF) crystal or in a glass. These Nd^{3+} lasers operate either pulsed or cw and lase at approximately 1064 nm. The high energies of pulsed Nd^{3+}:YAG lasers allow efficient frequency doubling (532 nm), tripling (355 nm), or quadrupling (266 nm), and the 532 nm and 355 nm beams are commonly used to pump tunable dye lasers.

Dye Lasers

The gain medium in a dye laser is an organic dye molecule that is dissolved in a solvent. The dye and solvent are circulated through a cell or a jet, and the dye molecules are excited by flashlamps or other lasers. Pulsed dye lasers use a cell and cw dye lasers typically use a jet. The organic dye molecules have broad fluorescence bands and dye lasers are typically tunable over 30 to 80 nm. Dyes exist to cover the near-uv to near-infrared spectral region: 330 - 1020 nm.

Schematic of a pulsed dye laser

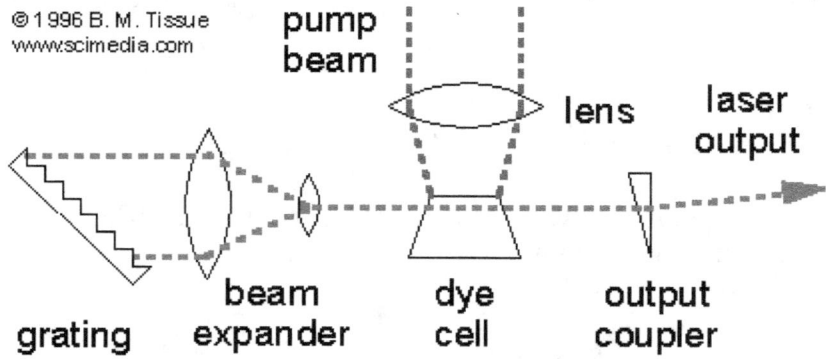

Semiconductor Lasers

Semiconductor lasers are light-emitting diodes within a resonator cavity that is formed either on the surfaces of the diode or externally. An electric current passing through the diode produces light emission when electrons and holes recombine at the p-n junction. Because of the small size of the active medium, the laser output is very divergent and requires special optics to produce a good beam shape.

Schematic of a semiconductor diode laser

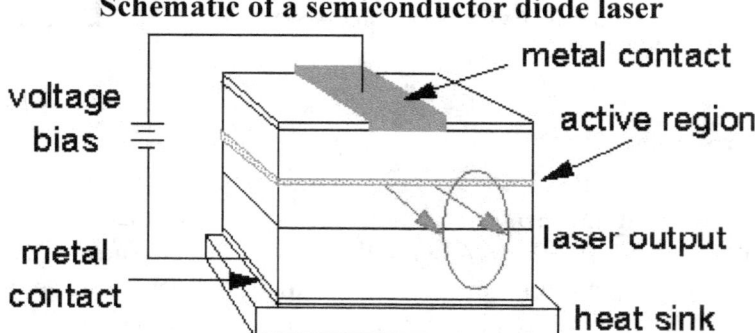

These lasers are used in optical-fiber communications, CD players, and in high-resolution molecular spectroscopy in the near-infrared. Diode laser arrays can replace flash lamps to efficiently pump solid-state lasers. Diode lasers are tunable over a narrow range and different semiconductor materials are used to make lasers at 680, 800, 1300, and 1500 nm.

Wavelength separators
Filters
Filters separate different parts of the electromagnetic spectrum by absorbing or reflecting certain wavelengths and transmitting other wavelengths.

Color Filters
Color filters are glass substrates containing absorbing species that absorb certain wavelengths. A typical example is a cut-on color filter, which blocks short wavelength light such as an excitation source, and transmits longer wavelength light such as fluorescence that reaches a detector.

Interference Filters: Interference filters are made of multiple dielectric thin films on a substrate. They use interference to selectively transmit or reflect a certain range of wavelengths. A typical example is a bandpass interference filter that transmits a narrow range of wavelengths, and can isolate a single emission line from a discharge lamp.

ptical Spectrometers: A spectrometer is an optical system that transmits a specific band of electromagnetic spectrum. Dispersion of different wavelengths is accomplished with the separating capability of refraction (prism) or diffraction (diffraction grating). Typical applications are isolation of a narrow band of radiation from a continuum light source for absorption measurements, or analysis of the emission from excited atoms or molecules.

Monochromator designs: A typical monochromator design is shown below. It consists of the diffraction grating (dispersing element), slits, and spherical mirrors.

Monochromator parameters: Bandpass : The wavelength range that the monochromator transmits.

Dispersion : The wavelength dispersing power, usually given as spectral range / slit width (nm/mm). Dispersion depends on the focal length, grating resolving power, and the grating **order.**

Schematic of a Czerny-Turner monochromator

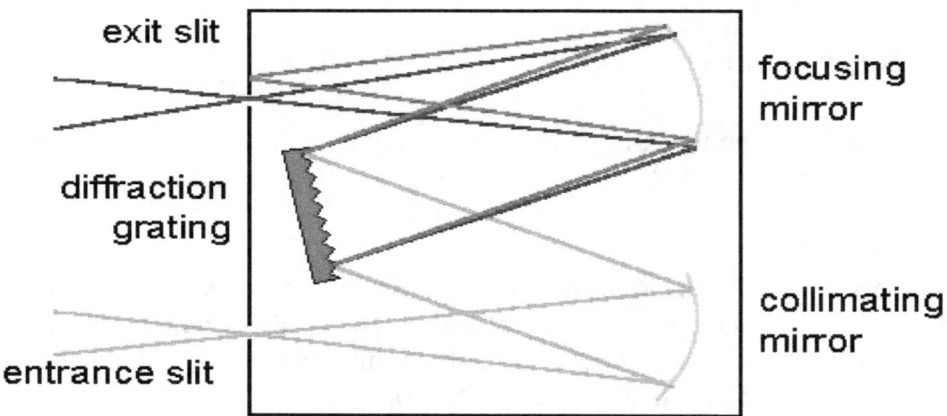

Scanning is accomplished by rotating the grating.

Resolution : The minimum bandpass of the spectrometer, usually determined by the aberrations of the optical system. Acceptance angle (f/#) A measure of light collecting ability, focal length / mirror diameter
Blaze wavelength : The wavelength of maximum intensity in first order.

Related instruments: Spectrograph : A spectrometer that records a wide bandpass with a photographic plate or an array detector. The spectrometer requires a flat image field.
Polychromator : A spectrometer with multiple detectors for simultaneous detection of multiple analytes.

Interferometers: The purpose of an interferometer is similar to that of a filter or monochromator, i.e., to isolate a specific portion of the electromagnetic spectrum. Unlike prism or grating monochromators, interferometers are not dispersive instruments, but use interference to selectively transmit a certain wavelength. The links below lead to descriptions of three interferometer designs.

Types of interferometers: Fabry-Perot interferometer (etalon): Used in high-resolution applications, such as atomic spectroscopy or measurement of narrow-band laser linewidths.
Michelson interferometer: Used in fourier-transform infrared absorption spectrometers (FTIR).
Mach-Zender interferometer: Used to measure refractive index changes in gases and in interference microscopes to image transparent samples.

Fabry-Perot Interferometer: The Fabry-Perot interferometer is commonly used as a narrow-bandpass filter or as an instrument to measure spectral linewidths.

Fabry-Perot interferometer (etalon)

Fabry-Perot interferometers that cannot be scanned are called etalons.

Schematic of a Fabry-Perot etalon

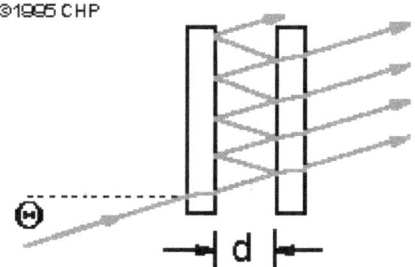

Transmission through a Fabry-Perot interferometer as a function of wavelength

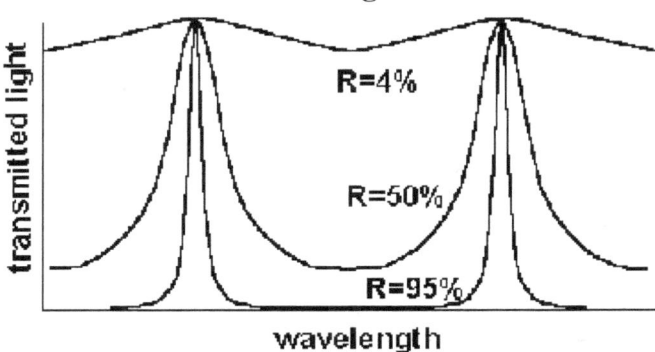

Michelson Interferometer

The Michelson interferometer design is used in Fourier-transform infrared absorption spectrometers (FTIR).

Schematic of a Michelson interferometer

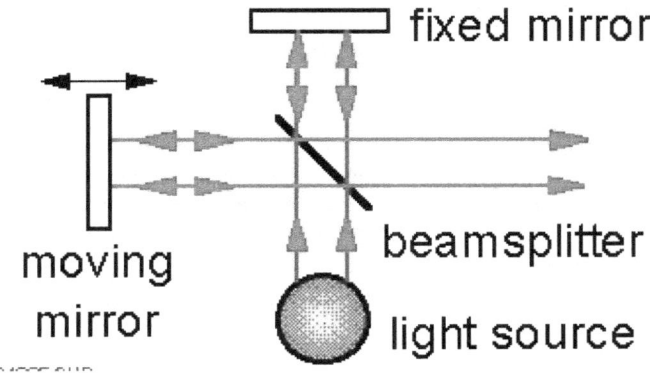

Optical Radiation Detectors

Detectors convert light energy to an electrical signal. In spectroscopy, they are typically placed after a wavelength separator to detect a selected wavelength of light. Different types of detectors are sensitive in different parts of the electromagnetic spectrum.

Mach-Zender Interferometer

The Mach-Zender interferometer design is used to measure refractive index changes in gases, and in interference microscopes to image transparent samples.

Schematic of a Mach-Zender interferometer

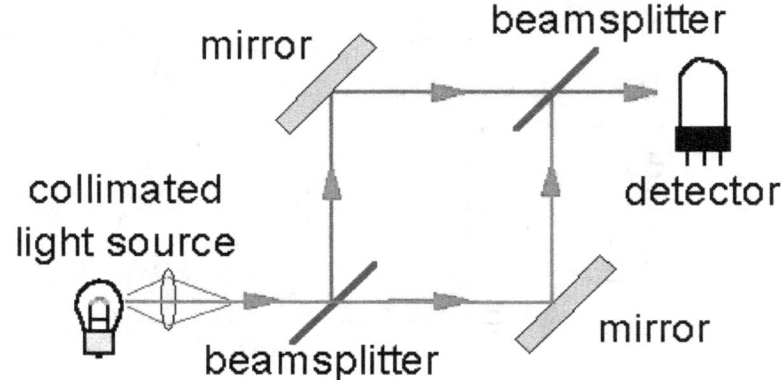

Charge-Coupled Devices (CCD):
A CCD is an integrated-circuit chip that contains an array of capacitors that store charge when light creates e-hole pairs.

Schematic of a CCD

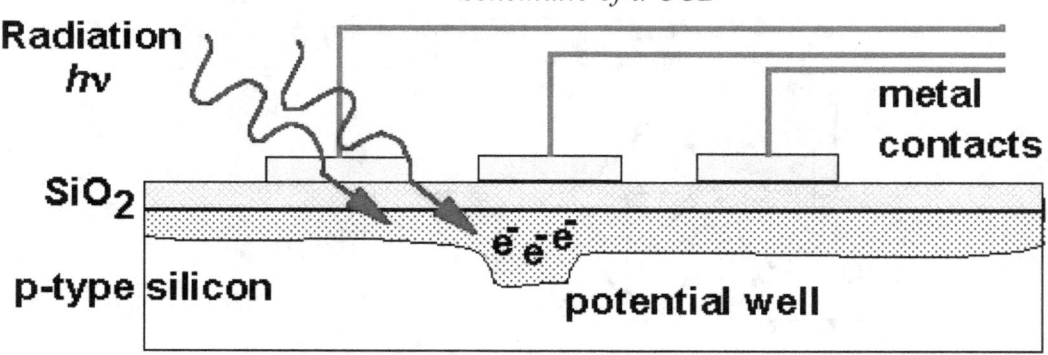

The charge accumulates and is read in a fixed time interval. CCDs are used in similar applications to other array detectors such as photodiode arrays, although the CCD is much more sensitive for measurement of low light levels.

Photodiode and Photovoltaic Detectors:

When a photon strikes a semiconductor, it can promote an electron from the valence band (filled orbitals) to the conduction band (unfilled orbitals) creating an electron(-) - hole(+) pair. The concentration of these electron-hole pairs is dependent on the amount of light striking the semiconductor, making the semiconductor suitable as an optical detector. There are two ways to monitor the concentration of electron-hole pairs. In photodiodes, a voltage bias is present and the concentration of light-induced electron-hole pairs determines the current through semiconductor. Photovoltaic detectors contain a p-n junction, that causes the electron-hole pairs to separate to produce a voltage that can be measured.

Schematic of semiconductor detector

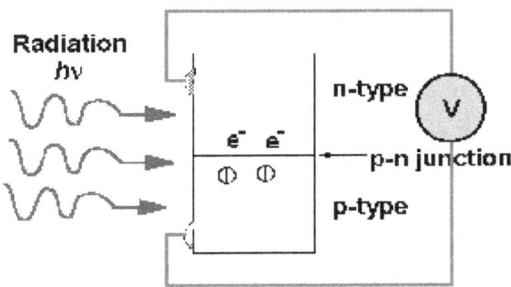

Photodiode Array Detectors (PDA)

A photodiode array (PDA) is a linear array of discrete photodiodes on an integrated circuit (IC) chip. For spectroscopy it is placed at the image plane of a spectrometer to allow a range of wavelengths to be detected simultaneously. In this regard it can be thought of as an electronic version of photographic film. Array detectors are especially useful for recording the full UV-VIS absorption spectra of samples that are rapidly passing through a sample flow cell, such as in an HPLC detector.

Schematic of a PDA

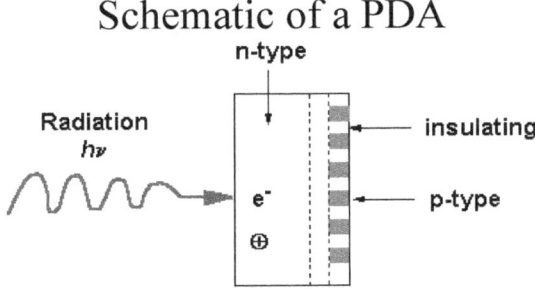

Light creates electron-hole pairs and the electrons migrate to the nearest PIN junction. After a fixed integration time the charge at each element is sequentially read with solid-state circuitry to generate the detector response as a function of linear distance along the array. PDAs are available with 512, 1024, or 2048 elements with typical dimensions of ~ 25 μm wide and 1-2 mm high.

Photodiode detectors are not as sensitive as PMTs but they are small and robust. Wavelength range ;

Detector type	Lambda
Si	0.2 - 1.1
Ge	0.4 - 1.8
InAs	1.0 - 3.8
InSb	1.0 - 7.0
InSb (77K)	1.0 - 5.6
HgCdTe (77K)	1.0 -25.0

Sensors:
A sensor is a device that produces a measurable response to a change in a physical condition, such as temperature or thermal conductivity, or to a change in chemical concentration. Sensors are particularly useful for making in-situ measurements such as in industrial process control or medical applications. A sensor is usually packaged as a complete unit, discrete light detectors and ion detectors are described in separate documents.

Ion-Selective Electrodes (ISE):
An Ion-Selective Electrode (ISE) produces a potential that is proportional to the concentration of an analyte. Making measurements with an ISE is therefore a form of potentiometry. The most common ISE is the pH electrode, which contains a thin glass membrane that responds to the H^+ concentration in a solution.

Theory: The potential difference across an ion-sensitive membrane is:

$$E = K - (2.303RT/nF) \log(a)$$

where K is a constant to account for all other potentials, R is the gas constant, T is temperature, n is the number of electrons transferred, F is Faraday's constant, and a is the activity of the analyte ion. A plot of measured potential versus $\log(a)$ will therefore give a straight line.

ISEs are susceptible to several interferences. Samples and standards are therefore diluted 1:1 with total ionic strength adjuster and buffer (TISAB). The TISAB consists of 1 M NaCl to adjust the ionic strength, acetic acid/acetate buffer to control pH, and a metal complexing agent.

Instrumentation
ISEs consist of the ion-selective membrane, an internal reference electrode, an external reference electrode, and a voltmeter. A typical meter is shown in the document on the pH meter.

Schematic of an ISE measurement

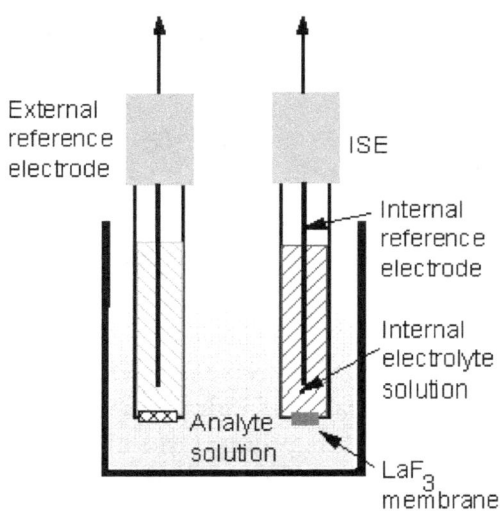

Commercial ISEs often combine the two electrodes into one unit that are then attached to a pH meter.

A commercial fluoride ISE

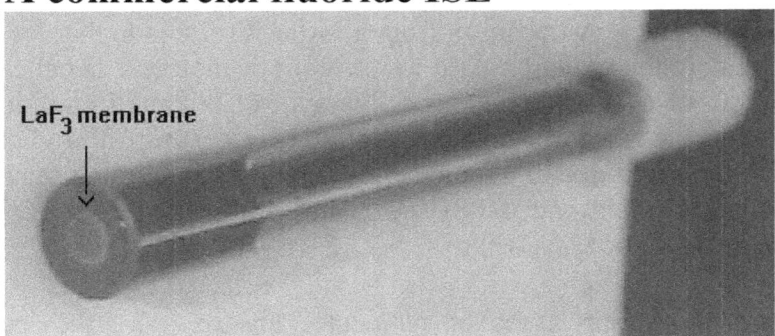

Thermal Conductivity Detectors (TCD): A TCD detector consists of an electrically-heated wire or thermistor. The temperature of the sensing element depends on the thermal conductivity of the gas flowing around it. Changes in thermal conductivity, such as when organic molecules displace some of the carrier gas, cause a temperature rise in the element which is sensed as a change in resistance. The TCD is not as sensitive as other dectectors but it is non-specific and non-destructive.

Instrumentation: Two pairs of TCDs are used in gas chromatographs. One pair is placed in the column effluent to detect the separated components as they leave the column, and another pair is placed before the injector or in a separate reference column. The resistances of the two sets of pairs are then arranged in a bridge circuit.

Schematic of a bridge circuit for TCD detection

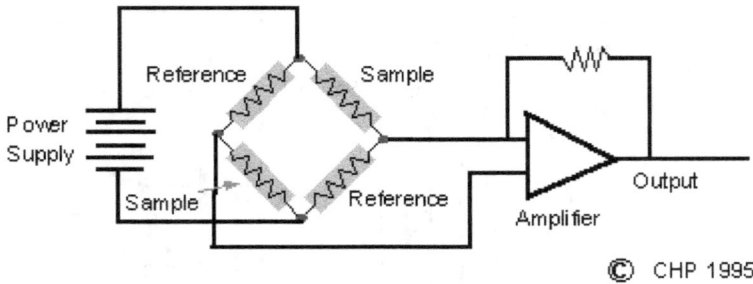

© CHP 1995

The bridge circuit allows amplification of resistance changes due to analytes passing over the sample thermoconductors and does not amplify changes in resistance that both sets of detectors produce due to flow rate fluctuations, etc.

Spectroscopy: Spectroscopy is the use of the absorption, emission, or scattering of electromagnetic radiation by atoms or molecules (or atomic or molecular ions) to qualitatively or quantitatively study the atoms or molecules, or to study physical processes. The interaction of radiation with matter can cause redirection of the radiation and/or transitions between the energy levels of the atoms or molecules. A transition from a lower level to a higher level with transfer of energy from the radiation field to the atom or molecule is called absorption. A transition from a higher level to a lower level is called emission if energy is transfered to the radiation field, or nonradiative decay if no radiation is emitted. Redirection of light due to its interaction with matter is called scattering, and may or may not occur with transfer of energy, i.e., the scattered radiation has a slightly different or the same wavelength.

Absorption: When atoms or molecules absorb light, the incoming energy excites a quantized structure to a higher energy level. The type of excitation depends on the wavelength of the light. Electrons are promoted to higher orbitals by ultraviolet or visible light, vibrations are excited by infrared light, and rotations are excited by microwaves.

An absorption spectrum is the absorption of light as a function of wavelength. The spectrum of an atom or molecule depends on its energy level structure, and absorption spectra are useful for identifying of compounds.

Measuring the concentration of an absorbing species in a sample is accomplished by applying the Beer-Lambert Law.

Emission: Atoms or molecules that are excited to high energy levels can decay to lower levels by emitting radiation (emission or luminescence). For atoms excited by a high-temperature energy source this light emission is commonly called atomic or optical emission (see atomic-emission spectroscopy), and for atoms excited with light it is called atomic fluorescence (see atomic-fluorescence spectroscopy). For molecules it is called fluorescence if the transition is between states of the same spin and phosphorescence if the transition occurs between states of different spin.

The emission intensity of an emitting substance is linearly proportional to analyte concentration at low concentrations, and is useful for quantitating emitting species.

Scattering: When electromagnetic radiation passes through matter, most of the radiation continues in its original direction but a small fraction is scattered in other directions. Light that is scattered at the same wavelength as the incoming light is called Rayleigh scattering. Light that is scattered in transparent solids due to vibrations (phonons) is called Brillouin scattering. Brillouin scattering is typically shifted by 0.1 to 1 cm^{-1} from the incident light. Light that is scattered due to vibrations in molecules or optical phonons in solids is called Raman scattering. Raman scattered light is shifted by as much as 4000 cm^{-1} from the incident light.

Spectrophotometry

A Spectrophotometer

In physics, **spectrophotometry** is the quantifiable study of electromagnetic spectra. It is more specific than the general term electromagnetic spectroscopy in that spectrophotometry deals with visible light, near-ultraviolet, and near-infrared. Also, the term does not cover time-resolved spectroscopic techniques.

Spectrophotometry involves the use of a spectrophotometer. A spectrophotometer is a photometer (a device for measuring light intensity) that can measure intensity as a function of the color, or more specifically, the wavelength of light. Important features of spectrophotometers are spectral bandwidth and linear range of absorption measurement.

Perhaps the most common application of spectrophotometers is the measurement of light absorption, but they can be designed to measure diffuse or specular reflectance. Strictly, even the emission half of a luminescence instrument is a kind of spectrophotometer.

The use of spectrophotometers is not limited to studies in physics. They are also commonly used in other scientific fields such as chemistry, biochemistry, and molecular biology. They are widely used in many industries including printing and forensic examination.

Design of a Spectrophotometer:
There are two major classes of spectrophotometers; single beam and double beam. A double beam spectrophotometer compares the light intensity between two light paths, one path containing a reference sample and the other the test sample. A single beam spectrophotometer measures the relative light intensity of the beam before and after a test sample is inserted. Although comparison measurements from double beam instruments are easier and more stable, single beam instruments can have a larger dynamic range and are optically simpler and more compact.

Historically, spectrophotometers use a monochromator containing a diffraction grating to produce the analytical spectrum. There are also spectrophotometers that use arrays of photosensors. Especially for infrared spectrophotometers, there are spectrophotometers that use a Fourier transform technique to acquire the spectral information quicker in a technique called Fourier Transform Infra Red spectrophotometer.

In short, the sequence of events in a spectrophotometer is as follows:
1. The light source shines into a monochromator.
2. A particular output wavelength is selected and beamed at the sample.
3. The sample absorbs light.
4. The photodetector behind the sample responds to the light stimulus and outputs an analog electronic current which is converted to a usable format.
5. The numbers are either plotted straight away, or are fed to a computer to be manipulated (e.g. curve smoothing, baseline correction and coversion to absorbency, a log function of light transmittance through the sample)

UV and IR spectrophotometers:

Beer and Lambert's law: When monochromatic light falls upon a homogenous solution a portion of incident light transmits, a portion of light is absorbed and a portion is reflected. Lambert states that when a monochromatic light passes away through a transparent medium the rate of decrease in intensity with the thickness of the medium is proportional to the intensity of the light. Light absorption and the light transmission for monochromatic light is a function of the thickness of the absorbing layer. Concentration of the colored substance has the definite effect on the light transmission or absorption. i.e. the intensity of the beam of a monochromatic light decreases exponentially as the concentration increases arithmetically. We can find an equation

O.D. = εct when

O.D = Optical Density of light absorbed

ε   =□Molecular extinction coefficient

t   = thickness of the path

Connection to mains: Read carefully the installation instruction regarding voltage requirement and earthing. To connect the instrument to the main power line, ensure the voltage and fuse setting are correct for the local use, Plug the main cable into the power socket and confirm the earth contact.

Routine maintenance:

1. Keep the interior of the instrument dust free, Keep the sample compartment clean and wipe out any split chemicals. Moisture cannot be allowed to leak into the instrument
2. Switch off and disconnect the instrument for the power line
3. Wipe the outer surface with a lint free cloth soaked with detergent solution
4. Wipe the outer surface with a dampen cloth. Wipe the surface with a dry cloth,
   Isopropyl alcohol may used for final wiping

Replacement of tungsten halogen lamp:

1. Keep the spectrophotometer at disconnected and cool state.
2. Follow the instruction given in the operator's manual to Lamp access panel and cover.
3. Carefully remove the lamp assembly and pull the ceramic socket pin of the lamp holder.
4. Wear gloves to fit the Tungsten Halogen Lamp. Touch only the base of the lamp or the mounting plate.
5. Insert the pin of the lamp with a clear tissue. Secure the lamp assembly according to the operator's manual.
6. Reconnect lamp cover, and fit all the screw of the lamp access panel.
7. Use the spectrophotometer after calibration properly.

Replacement of deuterium lamp:

1. Keep the spectrophotometer at disconnected and cool state.
2. Follow the instruction given in the operator's manual to Lamp access panel and cover.
3. Carefully deuterium Lamp from the supply socket and pull the ceramic socket pin of the lamp holder.

4. Wear gloves to fit the Lamp. Touch only the base of the lamp or the mounting plate.
5. Release the lamp assembly following operators manual
6. Insert the pin of the new lamp with a clear tissue. Secure the lamp assembly according to the operator's manual. Reconnect lamp cover, and fit all the screw of the lamp access panel. Use the spectrophotometer after calibration properly.
7. The UV radiation from a deuterium lamp is harmful to the skin and eyes. Wear apron and eye protectors.

Wavelength accuracy and repeatability:
1. Set power on and allow the warm the spectrophotometer and set wavelength at 351 nm at abs mode under closed lid.
2. Press Zero, insert the Holmium filter in the cell holder and close the lid.
3. Read and record the absorbance
4. Remove the holmium filter and select the wavelength mode
5. Increase the wavelength by 1 nm and record the wavelengths 351 to 371 nm.
6. Hopefully the peak reading will be at 361 $\pm$ 1 nm
7. Repeat the same test with Didymium filter within the wavelength range 527-to 547 nm .The peak will be at 537 $\pm$ 2 nm.
8. Repeat the same test with Didymium filter within the wavelength range 797 to 817 nm .The peak will be at 807 $\pm$ 2 nm
9. Repeat the test thrice. The allowable variation of the peak must not be greater than 1 nm.

Stray Light:
1. Set power on and allow the warm the spectrophotometer and set wavelength at 340 nm. at abs mode under closed lid.Lift the lid, insert the Blanking plate in the cell holder to block the beam and close the lid and set the dark current pressing zero.
2. Remove the Blanking plate and close the lid and press Zero.Insert a 1A gauge attenuator into the cell holder and close the lid. And record the reading (A) and press zero, remove the attenuator.
3. Insert 50g/l the aqueous (Na NO$_2$) solution in the cell holder and record reading (B)
4. (A) + (B) should be greater than 3.0 A or Insert 10g/l the aqueous (NaI) solution in the cell holder and record reading (B) at 220 nm.
5. Again Attenuator reading at 220 nm (A) + (B) should be greater than 3.0 A

Photometric accuracy:
1. Set power on and allow the warm the spectrophotometer and set wavelength at 546 nm at abs mode under closed lid and press enter.

2. Lift the lid, insert the Blanking plate in the cell holder to block the beam and close the lid and set the dark current pressing zero.
3. Remove the Blanking plate and close the lid and press Zero to set dark current.
4. Insert a zero absorbance neutral density filter into the cell holder and close the lid and record the reading. And then press zero.
5. Insert 2A calibration filter standard and compare the display reading with the previous reading. Variation should not be higher than ± 0.5 %, 02 A within 10 seconds in any case.

Setting an wavelength:

1. Set power on and allow the warm the spectrophotometer and wavelength 0f 340nm appears in the monitor. Let us adjust the wavelength at 695 nm
2. Touch the edit mode and first figure (0) of 0340 starts blinking, press enter, the second figure (3) of 0340 will blink, edit this figure 3 to desired number (say 6), adjust 6 by pressing up/down and press enter..
3. Then the third figure (4) of 0340 will blink, edit the figure 4 to desired number (say 9), adjust 9 by pressing up/down and press enter.
4. Then the fourth figure (0) of 0340 will blink, edit the figure 0 to desired number (say 5), adjust 9 by pressing up/down and press enter.
5. By this way wavelength is set at 695 nm.

Reading of the sample:

1. By pressing up/down set the abs mode and take the cuvette having blank inside the cuvette holder, Close the lid and press Zero to set Zero
2. Take the cuvette having Test/ Standard sample inside the cuvette holder, close the lid and record absorbance. Calculate the concentration of the test sample by comparing absorbance with that of standard.

Calculation :

$$C = \frac{\text{Conc. of the Standard} \quad \times \quad \text{abs of the Test} \times \quad \text{Dilution factor}}{\text{Abs of the Standard}}$$

IR spectrophotometry

Spectrophotometers designed for the main infrared region are quite different because of the technical requirements of measurement in that region. One major factor is the type of photosensors that are available for different spectral regions, but infrared measurement is also challenging because virtually everything emits IR light as thermal radiation, especially at wavelengths beyond about 5 μm.

Another complication is that quite a few materials such as glass and plastic absorb infrared light, making it incompatible as an optical medium. Ideal optical materials are salts, which do not absorb strongly. Samples for IR spectrophotometry may be smeared between two discs of potassium bromide or ground with potassium bromide and pressed into a pellet. Where aqueous solutions are to be measured, insoluble silver chloride is used to construct the cell.

Mass Spectrometry (MS)

Mass spectrometers use the difference in mass-to-charge ratio (m/e) of ionized atoms or molecules to separate them from each other. Mass spectrometry is therefore useful for quantitation of atoms or molecules and also for determining chemical and structural information about molecules. Molecules have distinctive fragmentation patterns that provide structural information to identify structural components.

The general operation of a mass spectrometer is:
1. create gas-phase ions
2. separate the ions in space or time based on their mass-to-charge ratio
3. measure the quantity of ions of each mass-to-charge ratio

The ion separation power of a mass spectrometer is described by the resolution, which is defined as: $R = m\ /\ delta\ m,$

where m is the ion mass and delta m is the difference in mass between two resolvable peaks in a mass spectrum. E.g., a mass spectrometer with a resolution of 1000 can resolve an ion with a m/e of 100.0 from an ion with an m/e of 100.1.

In general a mass spectrometer consists of an ion source, a mass-selective analyzer, and an ion detector. The magnetic-sector, quadrupole, and time-of-flight designs also require extraction and acceleration ion optics to transfer ions from the source region into the mass analyzer. The details of mass analyzer designs are discussed in the individual documents listed below. Basic descriptions of sample introduction/ionization and ion detection are discussed in separate documents on ionization methods and ion detectors, respectively.

Mass Spectrometry Ionization Methods
Chemical ionization (CI)

CI uses a reagent ion to react with the analyte molecules to form ions by either a proton or hydride transfer:

$$MH + C_2H_5^+ \dashrightarrow MH_2^+ + C_2H_4$$
$$MH + C_2H_5^+ \dashrightarrow M^+ + C_2H_6$$

The reagent ions are produced by introducing a large excess of methane (relative to the analyte) into an electron impact (EI) ion source. Electron collisions produce CH_4^+ and CH_3^+ which further react with methane to form CH_5^+ and $C_2H_5^+$:

$$CH_4^+ + CH_4 \dashrightarrow CH_5^+ + CH_3$$
$$CH_3^+ + CH_4 \dashrightarrow C_2H_5^+ + H_2$$

Plasma and glow discharge

A plasma is a hot, partially-ionized gas that effectively excites and ionizes atoms. A glow discharge is a low-pressure plasma maintained between two electrodes. It is particularly effective at sputtering and ionizing materials from solid surfaces.

Electron impact (EI)

An EI source uses an electron beam, usually generated fron a tungsten filament, to ionize gas-phase atoms or molecules. An electron from the beam knocks an electron off analyte atoms or molecules to create ions.

Electrospray ionization (ESI)

The ESI source consists of a very fine needle and a series of skimmers. A sample solution is sprayed into the source chamber to form droplets. The droplets carry charge when the exit the capillary and as the solvent vaporizes the droplets disappear leaving highly charged analyte molecules. ESI is particularly useful for large biological molecules that are difficult to vaporize or ionize.

Fast-atom bombardment (FAB)

In FAB a high-energy beam of netural atoms, typically Xe or Ar, strikes a solid sample causing desoprtion and ionization. It is used for large biological molecules that are difficult to get into the gas phase. FAB causes little fragmentation and usually gives a large molecular ion peak, making it useful for molecular weight determination.

The atomic beam is produced by accelerating ions from an ion source though a charge-exchange cell. The ions pick up an electron in collisions with netural atoms to form a beam of high energy atoms.

Laser ionization (LIMS)

A laser pulse ablates material from the surface of a sample and creates a microplasma that ionizes some of the sample constituents.

Matrix-assisted laser desorption ionization (MALDI)

MALDI is a LIMS method of vaporizing and ionizing large biological molecules such as proteins or DNA fragments. The biological molecules are dispersed in a solid matrix such as nicotinic acid.

A UV laser pulse ablates the matrix which carries some of the large molecules into the gas phase in an ionized form so they can be extracted into a mass spectrometer.

Plasma-desorption ionization (PD)

Decay of $_{252}$Cf produces two fission fragments that travel in opposite directions. One fragment strikes the sample knocking out 1-10 analyte ions. The other fragment strikes a detector and triggers the start of data acquisition. This ionization method is especially useful for large biological molecules.

Resonance ionization (RIMS)

One or more laser beams are tuned in resonance to transistions of a gas-phase atom or molecule to promote it in a stepwise fashion above its ionization potential to create an ion. Solid samples must be vaporized by heating, sputtering, or laser ablation.

Secondary ionization (SIMS)

An ion beam; such as $^3He^+$, $^{16}O^+$, or $^{40}Ar^+$; is focused onto the surface of a sample and sputters material into the gas phase. Approximately 1% of the sputtered material comes off as ions.

Spark source

A spark source ionizes analytes in solid samples by pulsing an electric current across two electrodes. If the sample is a metal it can serve as one of the electrodes, otherwise it can be mixed with graphite and placed in a cup-shaped electrode

Thermal ionization (TIMS)

Thermal ionization is used for elemental or refractory materials. A sample is deposited on a ribbon of Pt, Re The ribbon is often coated with graphite to provide a reducing effect.

Ion Detectors

Channeltron

A channeltron is a horn-shaped continuous dynode structure that is coated on the inside with a electron emissive material. An ion striking the channeltron creates secondary electrons that have an avalanche effect to create more secondary electrons and finally a current pulse.

Daly detector

A Daly detector consists of a metal knob that emits secondary electrons when struck by an ion. The secondary electrons are accelerated onto a scintillator that produces light that is then detected by a photomultiplier tube.

Electron multiplier tube (EMT)

Electron multiplier tubes are similar in design to photomultiplier tubes. They consist of a series of biased dynodes that eject secondary electrons when they are struck by an ion. They therefore multiply the ion current and can be used in analog or digital mode.

Faraday cup

A Faraday cup is a metal cup that is placed in the path of the ion beam. It is attached to an electrometer, which measures the ion-beam current. Since a Faraday cup can only be used in an analog mode it is less sensitive than other detectors that are capable of operating in pulse-counting mode.

Microchannel plate

A microchannel plate consists of an array of glass capillaries (10-25 um inner diameter) that are coated on the inside with a electron-emissive material. The capillaries are biased at a high voltage and like the channeltron, an ion that strikes the inside wall one of the capillaries creates an avalanche of secondary electrons. This cascading effect creates a gain of 10^3 to 10^4 and produces a current pulse at the output. Microchannel plates (MCP) are also used as an intensifier for low-intensity light detection with array detectors.

Schematic of a microchannel plate

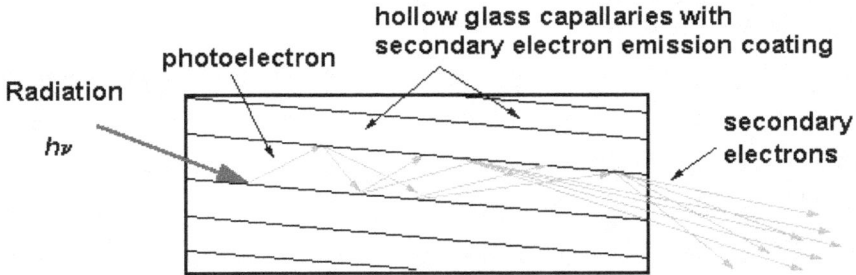

Ion-Trap Mass Spectrometry

The ion-trap mass spectrometer uses three electrodes to trap ions in a small volume. The mass analyzer consists of a ring electrode separating two hemispherical electrodes. A mass spectrum is obtained by changing the electrode voltages to eject the ions from the trap. The advantages of the ion-trap mass spectrometer include compact size, and the ability to trap and accumulate ions to increase the signal-to-noise ratio of a measurement.

Instrumentation

Single Focusing analyzers: A circular beam path of 180, 90, or 60 degrees can be used. The various forces influencing the particle separate ions with different mass-to-charge ratios.

Double Focusing analyzers: An electrostatic analyzer is added in this type of instrument to separate particles with difference in kinetic energies

Fourier-Transform Mass Spectrometry

Fourier-transform mass spectrometry takes advantage of ion-cyclotron resonance to select and detect ions.

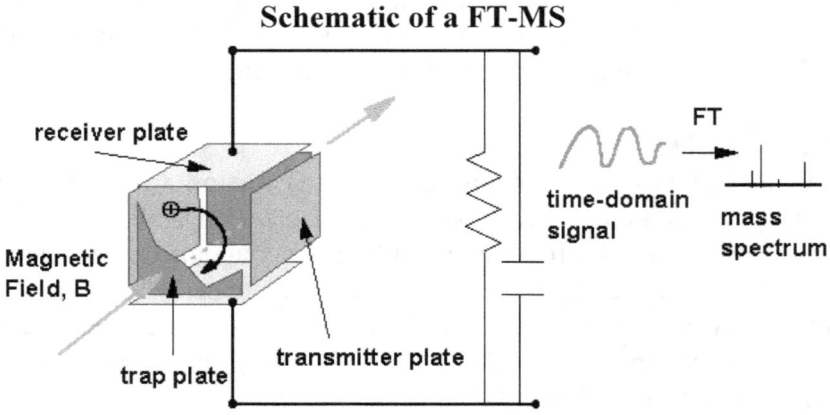

Schematic of a FT-MS

Magnetic-Sector Mass Spectrometry

Schematic of a magnetic-sector mass spectrometer

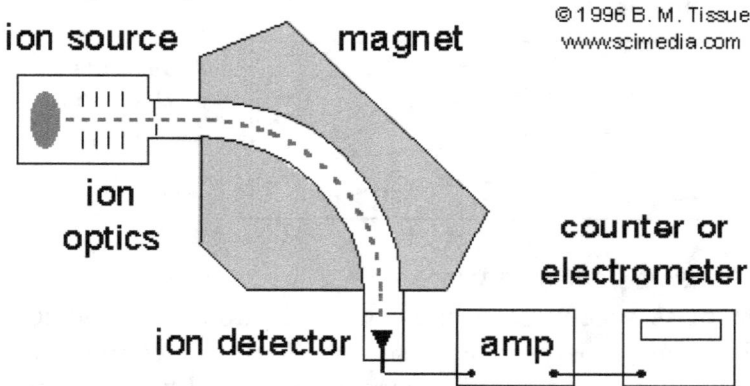

© 1996 B. M. Tissue
www.scimedia.com

Theory

The ion optics in the ion-source chamber of a mass spectrometer extract and accelerate ions to a kinetic energy given by: $K.E. = 0.5\, mv^2 = eV$
where m is the mass of the ion, v is it's velocity, e is the charge of the ion and V is the applied voltage of the ion optics.
The ions enter the flight tube between the poles of a magnet and are deflected by the magnetic field, H. Only ions of mass-to-charge ratio that have equal

centrifugal and centripetal forces pass through the flight tube: $mv^2 / r = Hev$
centrifugal = centripetal forces. Where r is the radius of curvature of the ion path:

$$\frac{m}{e} = \frac{H^2 r^2}{2V}$$

$r = mv / eH$, when

This equation shows that the m/e of the ions that reach the detector can be varied by changing either H or V.

Quadrupole Mass Spectrometry

A quadrupole mass filter consists of four parallel metal rods arranged as in the figure below. Two opposite rods have an applied potential of $(U+V\cos(wt))$ and the other two rods have a potential of $-(U+V\cos(wt))$, where U is a dc voltage and $V\cos(wt)$ is an ac voltage. The applied voltages affect the trajectory of ions traveling down the flight path centered between the four rods. For given dc and ac voltages, only ions of a certain mass-to-charge ratio pass through the quadrupole filter and all other ions are thrown out of their original path. A mass spectrum is obtained by monitoring the ions passing through the quadrupole filter as the voltages on the rods are varied. There are two methods: varying w and holding U and V constant, or varying U and V (U/V) fixed for a constant w. Quadrupole mass spectrometers consist of an ion source, ion optics to accelerate and focus the ions through an aperture into the quadrupole filter, the quadrupole filter itself with control voltage supplies, an exit aperture, an ion detector and electronics, and a high-vacuum system.

Time-of-Flight Mass Spectrometry (TOF-MS)

A time-of-flight mass spectrometer uses the differences in transit time through a drift region to separate ions of different masses. It operates in a pulsed mode so ions must be produced or extracted in pulses. An electric field accelerates all ions into a field-free drift region with a kinetic energy of qV, where q is the ion charge and V is the applied voltage. Since the ion kinetic energy is $0.5mv^2$, lighter ions have a higher velocity than heavier ions and reach the detector at the end of the drift region sooner.

Theory: K.E. = qV again $\frac{1}{2} mv^2 = qV$, $V = (^{2qV}/_m)^{1/2}$

The transit time (t) through the drift tube is L/V where L is the length of the drift tube.

$$t = L / (^{2V}/_{m/q})^{1/2}$$

Schematic of a quadrupole filter

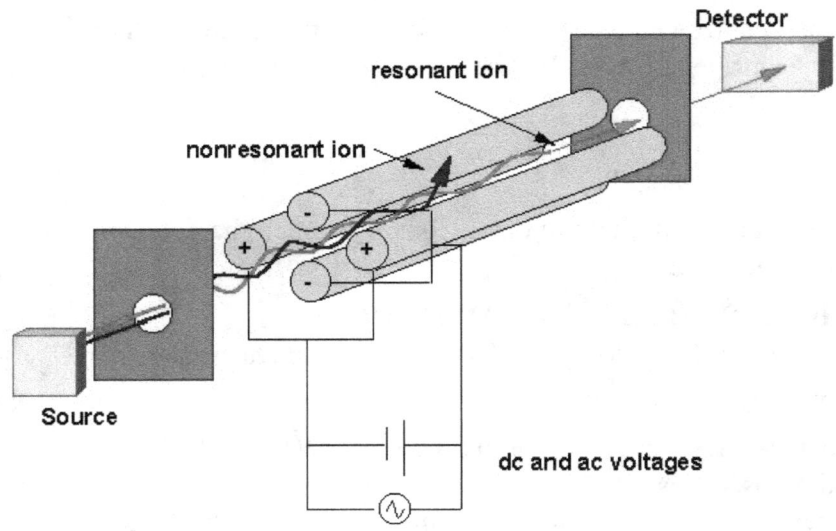

Example of a TOF mass spectrum

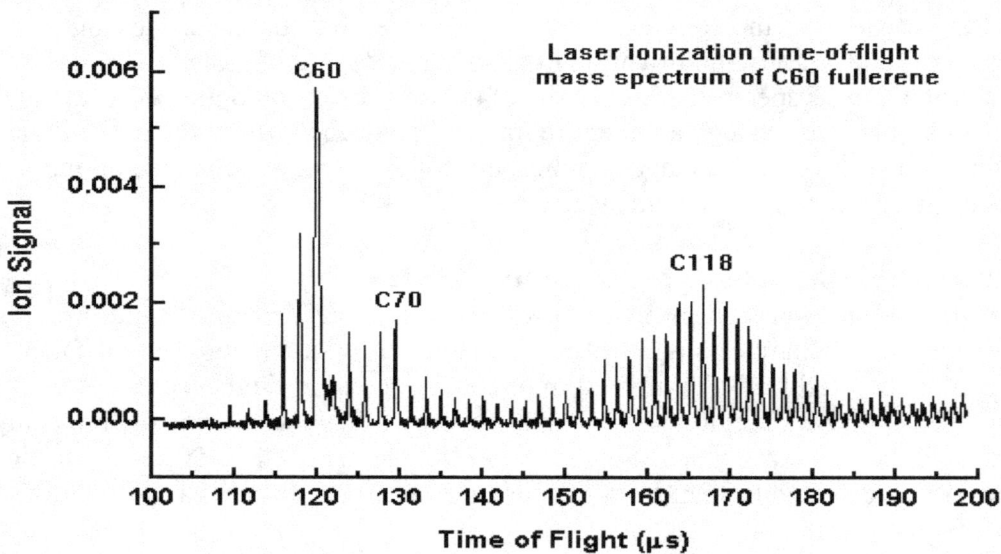

This schematic shows ablation of ions from a solid sample with a pulsed laser. The reflectron is a series of rings or grids that act as an ion mirror. This mirror compensates for the spread in kinetic energies of the ions as they enter the drift region and improves the resolution of the instrument. The output of an ion detector is displayed on an oscilloscope as a function of time to produce the mass spectrum

Schematic of a reflectron TOF-MS

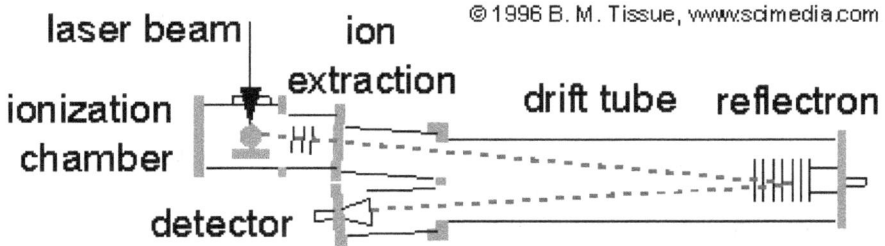

© 1996 B. M. Tissue, www.scimedia.com

Flame Photometric GC Detector

The reason to use more than one kind of detector for gas chromatography is to achieve selective and/or highly sensitive detection of specific compounds encountered in particular chromatographic analyses. The determination of sulfur or phosphorus containing compounds is the job of the flame photometric detector (FPD). This device uses the chemiluminescent reactions of these compounds in a hydrogen/ air flame as a source of analytical information that is relatively specific for substances containing these two kinds of atoms. The emitting species for sulfur compounds is excited S2. The $\square$max for emission of excited S2 is approximately 394 nm. The emitter for phosphorus compounds in the flame is excited HPO ($\square$max = doublet 510-526 nm). In order to selectively detect one or the other family of compounds as it elutes from the GC column, an interference filter is used between the flame and the photomultiplier tube (PMT) to isolate the appropriate emission band. The drawback here is that the filter must be exchanged between chromatographic runs if the other family of compounds is to be detected.

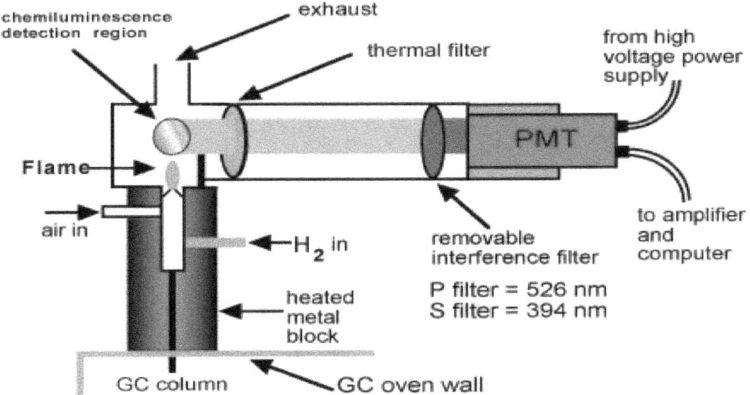

Instrumentation

In addition to the instrumental requirements for
1) a combustion chamber to house the flame,
2) gas lines for hydrogen (fuel) and air (oxidant), and

3) an exhaust chimney to remove combustion products, the final component necessary for this instrument is a thermal (band pass) filter to isolate only the visible and UV radiation emitted by the flame.

Without this the large amounts of infrared radiation emitted by the flame's combustion reactions would heat up the PMT and increase its background signal. The PMT is also physically insulated from the combustion chamber by using poorly (thermally) conducting metals to attach the PMT housing, Filters, etc. The physical arrangement of these components is as follows: flame (combustion) Chamber with exhaust, permanent thermal filter (two IR filters in some commercial designs), a removable phosphorus or sulfur selective filter, and finally the PMT.

5 LABORATORY SAFETY ISSUES

Laboratory Safety Issues

The responsibility for lab safety rests with each and every student in the laboratory. You must use common sense and work carefully to avoid chemical spills, broken glassware, and fires. This ensures not only your own safety, but that of your lab mates. Know the hazards of each chemical you use so that you will know what level of caution to use when handling it. If you do this, you will not be exposed to a harmful amount of any chemical during your year in organic chemistry lab.

If an accident does happen, you must take steps to prevent further injury. Most accidents are minor, and methods of dealing with them are detailed in the sections below. In the event of a serious accident, remember that injured people are often in shock and are unable to help themselves. You should be prepared to help your neighbor in case of an accident. A matter of seconds can be critical.

Chemical hygiene safety plan and links to OSHA regulations on laboratory safety

OSHA regulations (29 CFR) governing occupational exposure to hazardous chemicals in laboratories are linked below. These regulations require the adoption of a **Chemical Hygiene Plan (CHP)** for facilities using hazardous chemicals:

"Where hazardous chemicals as defined by this standard are used in the workplace, the employer shall develop and carry out the provisions of a written Chemical Hygiene Plan which is capable of protecting employees from health hazards associated with hazardous chemicals in that laboratory and capable of keeping exposures below the limits specified."

These regulations are currently non-mandatory in academic institutions. However, most colleges and universities are implementing CHP's. A CHP is written by the laboratory supervisors and addresses specific hazards in the laboratory and procedures for managing them.

Goggles and Eye Safety, Personal Protection Equipment, Special Health Problems: Goggles/Eye Safety

Researchers must wear chemical spill protection safety goggles whenever anyone in the lab room is handling chemicals. They must be flexible fitting, hood ventilation goggles, according to ANSI chemical splash standards. The CU bookstore sells this type of goggle. Past students rated several types of goggles and found the UVEX goggle (Stealth or Classic) as the most comfortable.

In case of spill a chemical in eye: If you get any chemical at all in your eye, immediately begin rinsing it in the eye wash, holding your eye open. Researcher/TA or another student will come to your assistance and help you ascertain the seriousness of the exposure.

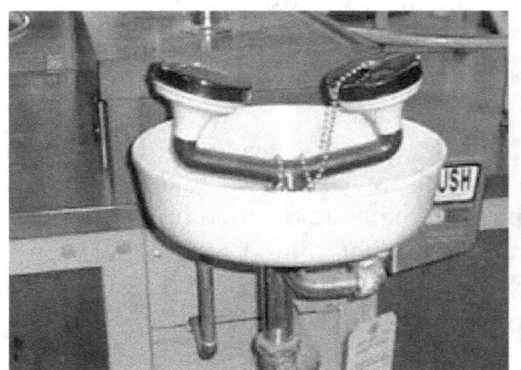

An eyewash in each lab room

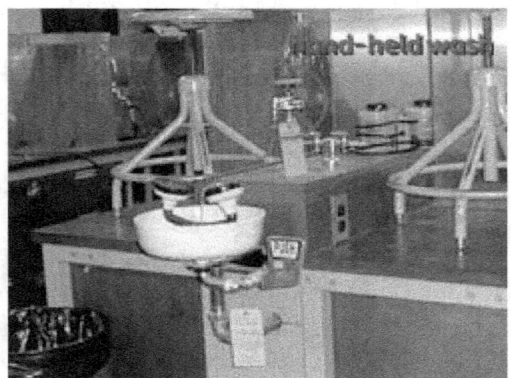

note the hand-held pull-out safety wash just above the eyewash

Personal Protection Equipment (PPE)

Personal protection equipment in the teaching laboratory includes safety goggles (discussed above), gloves, and lab coats and aprons. In industrial situations, PPE might include full suits of protective clothing, including boots and a face/head shield. Respirators are used when handling very toxic chemicals. If you are required to use a respirator, you will need to be trained in its use.

Clothing: Wearing required personal protection equipment (PPE) in lab during experiment is mandatory for researchers of any stage and status. Analysts must be covered from the top of the shoulders to well below the knees. (A bare middle is right at the edge of the lab bench where it can come into contact with spilled chemicals.) Analyst's feet must be covered–sandals are not appropriate in the chemical lab. Very loose fitting garments, such as ties and wide sleeves, as well as long unrestricted hair pose a hazard and must be restrained.

Special health problems

If Analysts are aware that they have allergies to specific chemicals or drugs, or to UV light, or if they have asthma or other health problems, they may want to consult their doctor before resuming in lab task. Every analyst should feel free to discuss any questions he may have with the Laboratory Coordinator. And please note:

"Analyst/Researchers may not take Molecular Medicine lab tasks if she is pregnant".

Overview of how to handle flammable, volatile, health hazardous and corrosive chemicals

One goal of the chemistry lab staff is to safely handle organic chemicals. Not only is this necessary for an analyst to have a safe experience in the lab, it is useful knowledge for almost any job they have after promotion.

Chemical Hazard Information for a Specific Compound

This information is found on one of many printed or online sources or on the material data safety sheet.

General Guide for Handling Chemicals in the Laboratory

Knowing what the hazards are is one thing, and knowing how to handle chemicals with these hazards is another.

Flammable Chemicals (examples: diethyl ether, acetone, hexanes, ethanol, methanol)

The method for proper handling of these flammable chemicals depends on their flammability rating, as may be given by a number 4-0 in the red area of a label. The label rating for diethyl ether is "4" while acetone, methanol, ethanol, and hexanes are "3". Ether is extremely flammable and any spark or simply heat can ignite it. The other four solvents listed here will readily burn, but they are not as likely to spontaneously combust. (Ether is also quite volatile, and its vapors can travel quickly away from the immediate work area. This fact increases its flammability danger.)

Ether cannot be used in a lab that has an open flame anywhere in the room. Extra care should be taken not to spill any flammable solvent (especially ether) on a heating mantle or hot plate. Ether should be kept away from electrical outlets.

In case of fire:

- o If clothing catches fire, immediately let the analyst drop to the floor and roll to smother the flames and call for help.
- o If a compound or solvent catches on fire, quickly cover the flames with a piece of glassware
- o If it is feasible, use a fire extinguisher to put the fire out.
- o Do not put water on an organic chemical fire because it will only spread the fire.

- o If the fire is large, do not take chances: evacuate the lab and the building immediately and tell your TA or the Coordinator what has happened
- o If no one in authority is available, pull the fire alarm in the hallway
- o If no one in authority is available, call (a network) from a safe phone.
- o If the fire alarm sounds for any reason, leave the room immediately and exit the building.

The types and use of fire extinguishers is covered in a separate orgchem web page:

- **Volatile Chemicals** (examples: hexanes, acetone, methylene chloride, diethyl ether)

Diethyl ether and methylene chloride are the most volatile of the chemicals.If they are accidentally inhaled, they can cause irritation of the respiratory tract, intoxication, drowsiness, nausea, or even central nervous system depression. Note that diethyl ether presents a special problem because it is not only volatile, it is also extremely flammable. nHandle all of these chemicals under a hood.

Atypical safety hood

Safety hood whenever possible shall used for experimentation, especially when volatile chemicals are being handled. Volatile chemicals through the lab, shall be carried in a covered container.

In case of gas spell: Evacuate the area immediately - at least into the hallway. The Coordinator shall be informed regarding the nature of the gas/vapor.

Contact Health Hazards of Solvents and Organic Chemicals (examples: methanol, ethanol, hexanes, acetone, methylene chloride, diethyl ether) The health hazard of a chemical is designated by a number 4-0 in the blue area of a label. None of the chemicals used has a "4" rating; most are 1 or 2.

In the past, chemicals have been spilled by students and left where they were in the lab, especially by the balances. This could cause serious harm to another student, so be sure to clean up a chemical spill promptly

In case of spill a chemical :
Immediately rinse the affected area with lots of water. Use soap if you wish, but never try to "treat" the spill with another solvent or chemical unless directed to do so by your Coordinator. If the affected area remains more than slightly red after the rinsing period, seek medical attention.

Hazards of Corrosives (examples: hydrochloric acid, sulfuric acid, phosphoric acid, nitric acid, sodium hydroxide)
Strong acids and bases are used frequently in the organic chemistry teaching labs. At full strength, they have a health rating of "3", meaning that short exposure could cause serious injury. (As they are diluted, the health rating drops about one number in rating for each 1:10 dilution.) If spilled on your skins, they cause a chemical burn. They are very harmful to your eyes. If you breathe in a big whiff of vapors, you will feel a burning in your nasal and respiratory passages.
Always wear goggles, gloves, protective clothing, and shoes. The heavier style of Playtex gloves are recommended for use when handling corrosives.

In case of spill of corrosives: Immediately rinse the affected area with lots of water. Use soap needed, but another solvent or chemical unless safety confirmed.

Glassware Safety
Use common sense when handling glassware. Glassware shall be kept away from the edge of the bench top. Each piece of glassware shall be inspected for hairline or star cracks before using it. When doing a distillation, each piece of glassware shall be clamped securely. The broken glassware shall be placed in one of the "Broken Clean Glassware" containers located in the labs.

If your reaction requires the use of a heating mantle or steam bath, glassware or the clamps used to hold glassware can become hot enough to cause a thermal burn on your skin. Wear heavy gloves to prevent this. Cuts can be also prevented by wearing thick gloves, especially while washing glassware. Protect your feet by wearing closed-toed shoes, not only to protect your feet from dropped glassware, but to protect them from broken pieces of glass which may be on the floor from a previous lab section. Always wear your goggles to protect your eyes from flying broken glassware.

Equipment and Electrical Safety:

Heating mantle, hot plate, or steam bath, shall be used with extra care because clamped glassware may be hot enough to cause a thermal burn.Heavy duty gloves are recommended.

Electrical equipment (heating mantles, Variacs, stir motors, hot plates) shall be carefully used properly to prevent electrical shock. The cord or plug shall be checked regularly to make sure that it is not damaged or frayed. The plug from the socket shall be disconnected by pulling firmly on the plug. Electrical cords that are left running across the floor are a trip hazard. Tripping and falling is especially dangerous in a chemistry lab, when a analyst might be carrying a corrosive or flammable chemical. Always wrap the cords firmly around stir motors, hot plates, and Mel Temps before you return them to the shelves for storage.

In case of cut or burn:

In case burns the wound shall be immediately washed with large amounts of cool water. Direct pressure may be applied to stop the bleeding as necessary. In case of the profuse bleeding, the affected limb shall be kept elevated. Thermal burns are treated by covering the affected area with cool water or ice.

Hazards of Chemicals:

Chemicals in the organic lab can be flammable, volatile, health hazardous, and/or corrosive. In the FIQC lab analysts require to **know the hazards** of all the chemicals in the laboratory. First and foremost, they need to know these hazards so that they will know when it is critical to take precautions such as wearing protective clothing or keeping chemicals from flame.

Laboratory Safety: Waste Chemical Handling:

Hazardous waste must be collected and processed according to guidelines set by the US Environmental Protection Agency (EPA) or Department of Environment Protection.

The hazardous waste :

Since the laboratory in this center the researchers are producing both the biological and chemical waste that shall be disposed properly as per safety requirements by checking compatibilities and pre-calculating percent compositions of all the waste.

All hazardous chemical wastes are collected in the main hood of each lab room, an area that is a designated "Satellite Accumulation Area (SAA)". The waste shall be placed as directed in the procedure section of each experiment in the Lab Manual.

Waste receptacles include:
- **Aqueous Waste** carboy - the 5 gallon plastic carboy located in the main hood, labeled with blue tape
- **Organic Waste** carboy - the 5 gallon plastic carboy located in the main hood, labeled with red tape
- Recovery Jar - a bottle or jar in the main hood for the collection of product
- **Recycling Container**s - usually 1 gallon glass jugs, labeled with the name of the solvent
- Used Acetone - a one gallon brown-glass bottle in each main hood

The system of waste tally will not work unless each and every analyst follows this quite simple and quite explicit rule, since the wastes are reported according to what is *supposed* to be in the carboys.

Waste rules:

- The waste always put in the waste receptacle as directed at the end of each procedure section of each experiment in the Lab Manual.
- Do not put acetone into the 5 gallon waste carboys; it goes into the one gallon glass bottle labeled **used acetone**.
- If you rinse a flask with water, use as little water as possible and always put the rinses in the **aqueous** waste carboy.
- When using acetone or water to rinse chemicals from glassware, use as little as possible (keep in mind that as soon as these rinses are placed in the hazardous waste container, they themselves become hazardous waste, thus burdening the hazardous waste collection system).
- Follow to the letter any directions that ask you to place a recrystallization solvent in a special receptacle; often we are able to collect and redistill a chemical rather than processing it as waste. Do not place rinse water or rinse acetone in these containers. Read labels carefully on recycled solvent receptacles.
- Pay attention to all special recovery containers.
- Pay special attention to halogenated hydrocarbon waste receptacles. Do not place water rinses in the halogenated waste.
- Place the products that you synthesized in the recovery jars in the main hood. Do not place filter or weighing papers in the recovery jars; instead, scrape the chemical off the paper and place the paper in the white rectangular waste paper tray in the main hood.
- If you have a waste chemical whose disposal is not specified in the manual, ask the Lab Coordinator what you should do with it.

Miscellaneous:
Filter papers. A container is provided in each hood for wet filter papers. The papers do not harm the environment and can be placed in the regular trash once

any solvent has evaporated from them. Do not place product in the wet filter paper container!!! The coordinator likes to assume that analysts have followed directions and placed the products in the recovery jar and not in the paper receptacle. Thus, the contents of the containers are placed into the trash cans. Please, do not allow lab chemicals to be placed in a local dumps.

Used drying agents. Drying agents such as sodium sulfate, magnesium sulfate, and calcium chloride are not harmful to the environment. They can be put in the regular trash as soon as any solvent has evaporated from them. Place them in the "Used Drying Agent" receptacle in the main hood. Do not place product in this receptacle, since it is emptied into the regular trash when the solvents have evaporated.

Pasteur pipets: Yes, Pasteur pipets are intended to be disposable and are quite inexpensive. However, we often find them laying around the lab and in the trash receptacles full of chemicals, including strong acids. All trash receptacles in the lab, including broken clean glassware receptacles, are eventually emptied into the local dumps. You can easily rinse a Pasteur pipet and use it in a future experiment. If one breaks, you must rinse it before you can throw it into the broken clean glassware receptacle or into one of the containers labeled "used mp capillaries, microcaps, Pasteur pipets". If you are unable to clean it with the minimum amount of solvent, consult the lab coordinator.

Used TLC plates: A small plastic container labeled "used TLC plates" will be placed in the main hood when necessary.

Used TLC spotters: These go into one of the containers labeled "used mp capillaries, microcaps, Pasteur pipets". These tiny spotters are hard to see, but can become lodged in your skin like a splinter. Please, put them in the receptacle when done with them.

Used melting point capillaries: These go into one of the containers labeled "used mp capillaries, microcaps, Pasteur pipets".

Broken glassware: If the glassware is not contaminated with chemicals, it can be placed in the plastic bucket labeled "Broken Glass" that is near the trash can in each lab room. If it is contaminated with chemicals follow directions given.

Remember: Analyst should avoid generation of unnecessary waste. 10 mL of unnecessary generated waste in a carboy each week during the year, 110 liters or 6 x 5 gallon carboys are unnecessarily produced.

put chemicals down the drain, you should place as little waste as is necessary in the waste carboys:

- If you are doing a recrystallization, place both unused solvent and solvent from the recrystallization in the specially labeled recovery bottle provided.
- When neutralizing a solution with an acid or a base, analysts usually take more than they need to their workspace. This is okay, but if you do not use all the solution, place it in the recovery bottle provided. The solution can then be reused or neutralized and placed down the drain.
- Acetone is a great solvent for cleaning organic chemistry glassware, and it is inexpensive. However, please use as little as possible, because it is not inexpensive for the FIQC Labs to process as hazardous waste. Currently, it makes up about 40% of the waste reported in organic chemistry. Use acetone sparingly.
- Use a very, very small amount of water to rinse your glassware and put this rinse water into the aqueous waste container. Then you can rinse your glassware in the sink. Don't put a large volume of water-rinses into the carboy, as this needlessly turns good water into hazardous waste, and it throws off the waste tally for that carboy, because the tally for the carboys includes an allocation for only a small amount of water rinses of glassware.

Environmental Health and Occupational Safety Agency:

Each laboratory has two large 5 gallon plastic carboys in the main hood. The carboy with blue tape is labeled "Hazardous Waste - Aqueous Waste" and the carboy with red tape is labeled "Hazardous Waste - Organic Waste". Note that there are marks in indelible ink on the side of each carboy. The highest mark indicates the last volume at which the content of the carboy was"tallied".

Each week, a different experiment is carried out in the lab. Under each main hood is a tally sheet on a clipboard. This sheet lists in English the percentages of the components of the waste mixtures which the students produce in each experiment. There is a separate list for "Organic" and for "Aqueous". At the end of each experiment, the Coordinator checks the waste carboy and makes sure that it is capped, and then marks the level with a Sharpie pen. A ruler is used to measure the distance between this new mark and the one just below it. Each inch corresponds to 1.6L; this volume is then written on the appropriate tally sheet (aqueous or organic). The system only works if each student places the waste in the container as directed in the Lab Manual. This cannot be overstated.

Chapter 6 GENERAL GUIDELINES OF GOOD LABORATORY PRACTICE

Laboratory shall follow the management and quality requirements ISO standard 17025.

Organization and Administration: Documents describing its scope of work, defined objective, financial strength, interaction between management and technical staff.

Staffing Pattern and Direction: Number of staff, Educational background and experience in similar set up and job description of staff.

Facilities and Equipment: Space, materials, production line, storage and wastage, safety of environment data storage, communication facilities, equipment and maintenance, records maintaining Validations.

Policy and Procedures: Standard operating procedure for purchase, storage, formulary and production, transportation, all experiment, methods of in-process quality control and disposal of wastage in writing form are available for all staff.

Staff development and Education: Planned continuing education program, in-house as well as outside training activities. A system of appraisal of staff members should be in position.

Research, Publication and product development: The laboratory should have formal policy for Development program and research activities to develop new products and improve products in use on the basis of scientific advancement.

Design and Construction features: The NIH and UCLA monograph for Molecular medicine Laboratory states should be manufactured in premises that are suitable for these purpose of handling animal cell lines, virus and pathogenic bacteria.

a) Buildings must be located, designed, constructed, adapted and maintained to suit the operation to be carried out within them.
b) Building shall have adequate space for the orderly placement of equipment and materials to prevent mix-ups between different components, product containers, closures, labeling, in process materials or drug products if any and to prevent contamination.
c) Laboratory experiment should be performed in specifically defined areas of adequate size suitable for appropriate sampling, testing or examination for QC unit and holding unused components before disposition.
d) There shall be space for quarantine storage before release of finished products if any.

e) Floor, wall and ceiling should be smooth and free from cracks and permits easy cleaning and disinfections.

f) Tissue culture, microbiological and chemical production areas should be separated.

g) Adequate washing facilities shall be provided including hot and cold water, detergent, air driers and clean toilet facilities easily accessible to working areas.

Ventilation, air filtration, air heating and cooling: Adequate ventilation with equipment for adequate control over air pressure, microorganisms, dust, humidity and temperature shall be provided appropriate for the handling of drugs, immunologicals and biologicals. Air is filtered with 0.22 micron HEPA filter before it is sent to the experimental area and animal house. Positive pressure designed to prevent outer microorganisms to enter. Negative pressure designed to prevent inner pathogenic /non-pathogenic microorganisms and other harmful materials for leaving room.

Plumbing and water supply: Potable water shall be supplied under continuous positive pressure in a plumbing system free of any defects that could contribute to contamination. Drains shall be of adequate size and direct a sewer having measures to prevent back siphoning.

Sewage and refuge: Sewage trash and other effuse in and from the building and immediate premises shall be disposed of in a safe and sanitary manner.

Animal Care and tissue culture facilities: Seed lots, cell banks and animals used for the experiments should be separated from other areas, materials with restricted access. Cleaning procedures shall be in written form and in an assigned manner. There shall be written procedures to use rodenticides, insecticide, and fungicide, fumigating agents and cleaning agents.

Environmental monitoring: A clean zone is a defined space in which concentration of airborne particles is controlled to specified air borne particulate cleanliness class. It is necessary to assess the microbial levels in the air and on the surface of the clean rooms to control the bio contamination. Microbial levels of process water and gases, raw materials and cell lines be monitored in a regular basis as well as monitoring microbial levels of the personnel working area. All solutions and media used throughout the experimental process must be tested for microbial load.

A clean room environment is controlled environment to meet a specified cleanliness class in terms of airborne particles and microorganisms.

Class Name		Upper Limit for 0.3 μm particles	
Metric	US	Particles per m^3	Particles per m^3
M 1		10.0	0.283
M1.5	1	35.3	1
M2		100	2.83
M2.5	10	353	10
M3		1000	28.3
M3.5	100	3530	100
M4		10000	283
M4.5	1000	35300	1000
M5		100000	2830
M5.6	10000	353000	10000
M6		1000000	28300
M6.5	100000	3530000	100000
M7		1000000	283000

Table 1: US and International Clean Room Standard

a) Assessment of the microbial levels in the air and on the surfaces of clean room in order to confer bio contamination according to specification.

b) Microbial level of process water and liquids media must be determined on routine basis,

Classification	Total particles > 0.5 μm / m^3	Total particles > 0.5 μm / m^3	Cfus per m^3
A$^+$	3500	0	< 1
B	3500	0	5
C	350000	2000	100
D	3500000	2000	500

Table 2:Clean Room Standard According to European Pharmacopoeia

Personal Hygiene: All clothing worn under clean room garments must be clean, not frayed and nonlinting. Gowning is done from head down to prevent contamination from dropping from street clothes on to clean room particularly in the animal house, microbial laboratory. All excess personal items shall be stored in lockers. Items such as pens, keys, combs etc. should be left in the locker. Garment should be of perfect size. Large garments can create bagpipe effects.

Good personal laboratory practices are not inherited but are to be acquired upon training and logical thinking. These practices are applicable to regular staff, maintenance crew, visitors,

Activity	0.3 □m Particles released /min
Sitting or standing	100,000
Walking + 3.5 km/hr	5,000,000
Walking + 6.0 km/hr	7,500,000
Walking + 8.5 km/hr	10,000,000

Table 3: Particles released during various activities by humans

Training of the employees: Laboratory should have a defined and effective training program for the employees both internal and external. The internal training of the employees who regularly enter the production area should include the following aspects:

- Fundamentals of clean room design, operation and monitoring
- Proper behavior in the clean room and Safety measures in the laboratory
- Correct gowning procedures and Personal hygiene
- General principles of microbiology and Handling of products in the clean room

Processing records: Processing records of Molecular Medicine as a Biological laboratory must include the history of each lot of a reagents showing that it has prepared, tested, dispensed into containers and used according to a required procedure. This processing record contains the following information:

- The name and concentration of the reagent or media prepared with date of preparation ;
- The complete formulation of the lot, including identification of seed and starting material;
- The passage number of each cell culture used in the experiment;
- A duly signed record of each step followed, precautions taken and special observation;
- A record of all in process control tests and results

Environmental variations have a specific impact on the mental and physical health conditions of animals, thereby results in a misleading outcome of the study. This variation may occur in both the macro and micro environmental conditions, which should be controlled. Environments of an animal house should permit the following:

a) Most of the laboratory animals like rabbits, guinea pigs and mouse can tolerate a relative humidity of 40-70% and a temperature range of about 19° to 26° C. Normally this should be maintained or may be adjusted according to the animal we want to use.

b) A thermograph for recording temperature variations and a hygrometer for moisture variation should be installed.

c) Air of the animal house exchanged with fresh air free from any particulate matter at least 10-15 times an hour depending on the number of species and animals housed.

d) Ammonia control is very important, as the ammonia is the metabolic product of urea in excreted urine of the animals by bacterial contamination. Room should be well ventilated and the bedding materials of the cages should be changed accordingly.

e) Laminar airflow system combined with HEPA filter reduces the airborne infections on laboratory animals.

f) An adequate pre-HEPA filter can be installed to remove dust and hair produced by the caged animals with 12-hour light and dark cycling system.

g) Noise leveling the animal area should be kept minimum level. Otherwise at stressed condition of high noise level may cause enlargement of adrenal gland, reduce fertility, increased blood pressure, auditory damage and behavioral disorders.

h) Walls and floors must be free from cracks and crevices, pipelines, drains and air filters should be well sealed to inhibit vermin to enter.

i) Sticky traps may be placed in animal, feeding and service rooms to determine the entry of roaches and insects. The required measures can be taken to control their entry into the animal house.

j) Physical separation of animals by species ensures the prevention of interspecies transmission of disease. All animals should be regularly observed for signs of illness, injury or abnormal behavior.

k) Unexpected deaths and signs of illness, distress or other deviation formal animals should be recorded. Serological testing, bacterial culture, histopathology and DNA analysis by PCR may be used in combination to detect types of infections occurred.

l) Euthanasia may be carried out in a manner that avoids animals' distress or in absence of other animals. Each animal room should have an entry and an exit door. The entrance door leads from the clean corridor to the animal room and the exit door leads the dirty things to an exit dirty corridor.

Animal Care and Use Protocols: The following topics should be considered in the preparation and review of animal care and use protocols:
Rationale and purpose of the proposed use of animals.

- Justification of the species and number of animals requested. Whenever possible, the number of animals requested should be justified statistically.
- Availability or appropriateness of the use of less-invasive procedures, other species, isolated organ preparation, cell or tissue culture, or computer simulation
- Adequacy of training and experience of personnel in the procedures used.

- Unusual housing and husbandry requirements.
- Appropriate sedation, analgesia, and anesthesia could be applied and unnecessary duplication of experiments should be avoided.
- Conduct of multiple major operative procedures.
- Criteria and process for timely intervention, removal of animals from a study, or stressful outcomes are anticipated.
- Post procedure care should be ensured..
- Method of euthanasia or disposition of animal.
- Safety of working environment for personnel.

THE SCHEMATIC DIAGRAM OF AN ANIMAL HOUSE

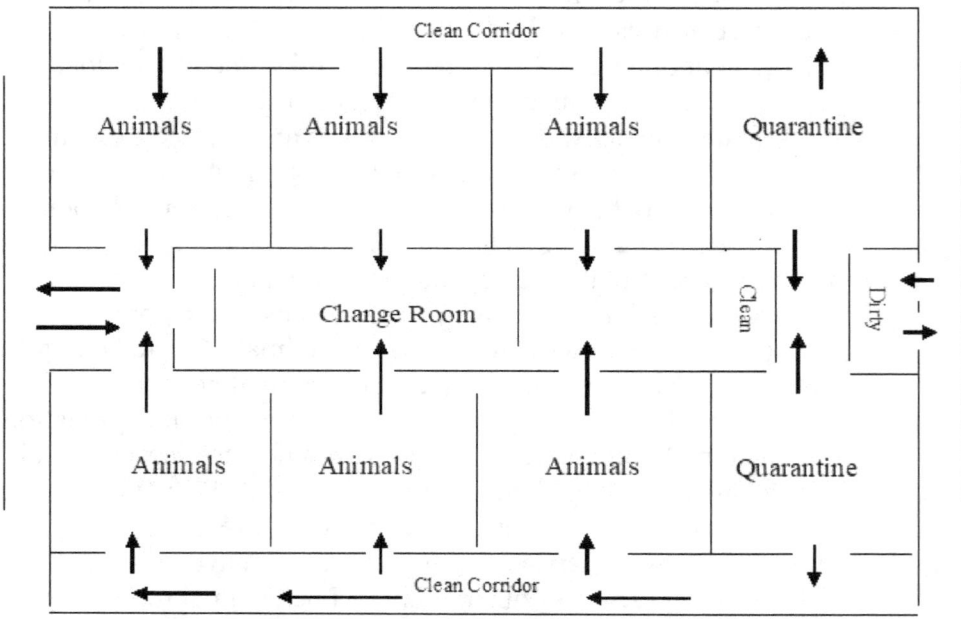

7 Master validation plan for a Molecular Medicine Laboratory

Introduction:
This master validation plan will be followed by the vaccine production facilities at IPH to get a classified quality air and a space for processing their products. This validation plan is designed to outline the necessary validation for the facilities including utilities and equipment used in the production.

Validation Stages :
The different stages of validation may be classified as different qualification protocols.
- Design Qualification (DQ)
- Room Qualification (RQ)
- Installation Qualification (IQ)
- Operational Qualification (OQ)
- Performance Qualification PQ)
- And Re-qualification RQ)

Design qualification (DQ) constitutes the assurance that the premises, utilities, equipment and processes have been designed according to the quality requirements.

Installation Qualification (IQ) is related to the performance tests to ensure that the installation machines, measuring devices, utilizes and areas used in the laboratory are
- Selected appropriately according to specification
- Correctly installed, and
- Operates accordingly.

Operational Qualification (OQ) is related to performance of the equipment to ensure that the function of installation machines, measuring devices, utilizes and areas used in the laboratory are being operating according to its operational specification.

Performance Qualification (PQ) is relatyed to the consistent performance of the installation machines, measuring devices, utilities according to required specification during routine operation.

Re-qualification (RQ) is related to calibration, verification and maintenance for periodic adaptation of equipment, maintenance, movement and repairs.

Responsibility:
The following departments are involved in the plan of validation:

a. Engineering staff for their renovation, installation and operation of an equipment
b. Production staff for in process sampling and submit them to QC
c. QC staff to test sample and report thereof

Quality Assurance Department to coordinate the documentation and develop validation package,

Qualification Report :
A written report should be available after completion of the validation or qualification includes:

- The title and objective of the study,
- Reference to the protocol,
- Details of material and equipment,
- Programmes and cycles used,
- Details of procedures and test methods
- A conclusion and recommendation as to the suitability of the equipment
- Duly approved and authorized (signed and dated).

Calibration and Verification: Regular calibration, validation and verification of all equipment, instruments and other devices used to measure the physical properties of substances, must be performed regularly according to accepted programme. Equipments should be listed, together with the following information for each piece of equipment calibration standards and limits, responsibilities for performing calibration, intervals between calibration, record keeping requirements and logs and actions to be taken when problems are identified.

After calibration each piece of equipment, instruments and other devices under the control of the laboratory, and requiring calibration, should be labelled, coded or otherwise identified to indicate the status of calibration and the date when re-calibration is due.

Master Validation Plan : This is the first document to be reviewed by during inspection by a regulatory or control authority.

- It is a formal policy document which describes the overall philosophy of the company towards validation and which also describes the key elements of the validation programme, organizational structure of validation, schedules and responsibilities. It should describe: "Why, what, where, by whom, how and when?".

- The VMP should direct to the more specific, detailed documents such as protocols, reports and documentation preparation and their control, SOPs, and personnel training records.

- The VMP should identify which systems, facilities, equipment and processes are subject to validation, the nature and extent of such testing and the applicable validation and qualification protocols and procedures.

Specific Requirements

The VMP should be concise and should typically include the following:

- Table of contents; introduction, policy and objectives, description of facilities, including plans;
- Constitution of the validation committee;
- Description and history of equipment and listing of protocols;
- Preventative maintenance programme and personnel training programme;
- Process and cleaning validation and key acceptance criteria;
- Laboratory instrument qualification and analytical method validation;
- Facility/utility qualification, computer system validation and re-validation intervals;
- Reasonable unexpected events (worst case), e.g. power failure, computer crash and recovery, filter integrity test failure;
- Documentation format to ensure a systematic approach to the layout and format of these documents, e.g. training record, raw data retention record, calibration record, validation protocol, validation report, etc.;
- Planning and scheduling ;
- Location where the validation activity is to be performed
- Estimate of staffing requirements to complete the validation effort described in the plan
- Time plan for the project, showing detailed planning of sub-projects ;
- Change control identifying the company's commitment to controlling critical changes and
  approvals and documentation

Validation Protocol

- The VP should include at least significant background information, the objectives of the validation and qualification study, site of the study, the responsible personnel, description of SOPs to be followed, equipment (including calibration before and after validation), standards and criteria for the relevant products and processes, the type of validation, and frequency.

- The processes and/or parameters to be validated (e.g. mixing times, drying temperatures, particle size, drying times, physical characteristics, content uniformity, etc.) should be clearly identified.

- Pre-determined acceptance criteria for drawing conclusions should be provided, as well as a description on how the results will be analyzed.

Validation Report
- The report should include the title and objective of the study, and should refer to the protocol, details of material, equipment, programs and cycles used, procedures and test methods.

- Recommendations on the limits and criteria to be applied to all future production batches, which should form part of the basis of the future batch manufacturing document, should be included.

- The results should be evaluated, analysed and compared with the acceptance criteria. All results should meet the criteria of acceptance and satisfy the stated objective. If necessary further studies should be performed.

- If found acceptable the report should be approved and authorized (signed and dated).

Supporting Facilities
Supporting facilities should be subjected to the different stages of qualification (e.g. Design Qualification (DQ), Installation Qualification (IQ), Operational Qualification (OQ) and Performance Qualification (PQ)) and should be recorded.
Supporting facilities/systems include waste systems, e.g. process drain systems, hazardous waste, solid waste disposal systems, air handling systems such as Heating Ventilation and Air Conditioning (HVAC), air filtration, laminar flow hoods, water systems such as RO water; gas systems such as compressed air, gas supply (nitrogen, oxygen and other gases), and electrical systems such as electrical emergency power and back-up power.

Heating, Ventilation and Air Conditioning (HVAC) System
The HVAC system plays an important role in product protection, personnel protection and environmental protection.

For all HVAC installation components, sub-systems or parameters, critical parameters and non-critical parameters should be determined. Some of the typical HVAC system parameters that should be qualified include:
- Room temperature and humidity;
- Supply air and return air quantities;

- Room pressure, air change rate, flow patterns, particle count and clean-up rates; and
- Laminar flow velocities and HEPA filter penetration tests

A. Utilities Validations: Heating, Ventilation and Air Conditioning (HVAC) system

1. Installation: For installation of HVAC system the given instruction should be followed properly and any discrepancies will be noted. All utilities operable under HVAC system will be monitored for proper functioning. Documentation will be done for installation, testing, operation, maintenance manual and spare parts list.
2. Operation: The system is said to be operated only when the following parameters are achieved within the specification:
   a) Air differential pressure
   b) Air changes
   c) Temperature control
   d) Viable and non-viable air counts
3. Performance: Performance testing will include non-viable monitoring in every month and one viable air test per day (rooms will be rotated on a daily basis) for four weeks.
4. Maintenance, calibration and audit program:
   a) From the daily monitoring of viable and nonviable air counts during normal activity in the air, an annual bacteriological profile may be prepared. Any critical processes will be monitored each time.
   b) HVAC monitoring is performed every six months.
   c) Air pressure will be monitored daily to ensure correct airflow.
   d) This HVAC system will be considered as preventive maintenance program

B) Utilities validation system: Compressed process air system

1. Installation: The installation of piping and process connections will be according to given instruction and design requirements and any discrepancies will be noted. Documentation will be done for installation, testing, operation, maintenance manual and spare parts list.
2. Operation: The Air differential pressure will be monitored during peak equipment
   use to ensure optimum capacity. The alarm will be tested for proper functionality.
3. Performance: Performance testing will include monitoring on air pressure and purity of air supply (oil free and no detectable hydrocarbons).
4. Maintenance, calibration and audit program:
a) Daily monitoring for reliable system and proper mechanical operation.

b) A logbook for maintenance and monitoring will be practiced.

C) Equipment validation: Classed Cold Room:

1.Installation: For installation of Cold Room the given instruction should be followed properly and any discrepancies will be noted. The system will be monitored for proper functioning. Documentation will be done for installation, testing, operation, maintenance manual and spare parts list.

2. Operation: The system is said to be operated only when the following parameters are achieved within the specification:

   (b) Air differential pressure,
   (c) Air changes
   (d) Temperature control and
   (e) Viable and non-viable air counts

3. Performance: Performance testing will include non-viable monitoring and one viable air test per day (rooms will be rotated on a daily basis) for four weeks.

4. Maintenance, calibration and audit program:

   a) From the daily monitoring of viable and nonviable air counts during normal activity, an annual bacteriological profile may be prepared. Any critical processes will be monitored each time.
   b) HEPA monitoring is performed every six months.
   c) Air pressure will be monitored daily to ensure correct air flow
   d) This cold room system will be considered as preventive maintenance program

D) Equipment validation: Unclassed Cold Room and incubators:

1) Installation: For installation of Cold Rooms and incubators, instruction should be followed properly and any discrepancies will be noted. All utilities that are required for the correct operation will be monitored to ensure they are within the specified range. Documentation will be done for installation, testing, operation, maintenance manual and spare parts list.

2) Operation: The cold rooms and incubator for temperature achieved within the specification. Temperature alarms and safety features will be tested for proper functioning.

3) Performance: The performance of the cold rooms and incubators include a continuous recording chart and personal recording internal temperature with certified thermometer.

4) Maintenance, calibration and audit program: There will be continuous recording chart on each cold room with daily monitoring by personnel. The cold rooms and incubators would be maintained quarterly.

E) Equipment validation: Biological Safety Cabinets and horizontal Hoods and Fume Hoods:

1) Installation: For installation of Cold Room the given instruction should be followed properly and any discrepancies will be noted. All utilities that are required for the correct operation of the hood will be monitored to ensure that are within the specified range. Documentation will be done for installation, testing, operation, maintenance manual and spare parts list.

2) Operational and performance Qualifications: At a minimum, this qualification will include air velocity profiles, filter seal and filter integrity testing, UV light, and laminar flow system.

3) Maintenance, Calibration Audit program: A maintenance, cleaning and use log will be included for each hood. A cleaning validation will be performed for the campaign cleaning of the toxoid purification room. All critical processes in any hood will be monitored with non-viable air and viable surface testing.

Typical laboratory equipment sanitizing schedule

Equipment	Sanitary frequency
Shoebox cages	Once or twice a week
Hanging wire cages or racks	Every two weeks
Removable feeders and feed buckets	Once in two days Once a day
Water bottles	Daily rinsing, Twice weekly sanitation
Water bowls	High pressure flush or chemical treatment
Automatic watering device	once weekly
Trash cans	Twice a month
Pens or runs	Daily with extensive cleaning every week

Water System Validation :

All water-treatment systems should be subject to planned maintenance, validation and monitoring.

Validation of water systems should consist of at least three phases:
- **Phase 1:** Investigational phaseDuring the first 2-4 weeks of the commissioning

of the plant the DQ, IQ and OQ should be performed. Operational parameters should be established and the cleaning and sanitization procedures, including frequencies for cleaning and sanitization, should be determined.

Daily sampling should be performed at each point of use. On completion of the phase the SOP for the water system should be developed.

- **Phase 2:**Short-term control

During the following 4-5 weeks the control of the system should be verified.Sampling should be performed as during Phase 1.

- **Phase 3:** Long-term control

During the following year the objective should be to demonstrate that the system is in control over a long period of time. Sampling may be reduced to weekly. The validation performed and re-validation requirements should be included in the Water Quality Manual.

Equipment:

Requirement for qualification should be applied to equipment used in production as well as in quality control laboratories.

The Design Qualification (DQ) should define the functional and operational specifications of the instrument and should detail the conscious decisions in the selection of the supplier.

Prior to use and to ensure that the equipment is fit for its intended use the different stages of qualification should be performed, e.g. IQ, OQ, and PQ.

In addition, the equipment should be well maintained and routinely calibrated. Certain stages of the equipment qualification may be done by the supplier or a third party.

Each major piece of equipment should have a logbook, which should detail at least, the supplier's name, module, model and serial number, date of installation, all qualification performed, maintenance performed and reference to records, and routine use.

Processes: Production processes should be validated. Process validation should only begin when qualification is complete.

Process validation should be organized and administered in the same way as qualification. It should be associated with the writing and issuing of process validation protocols, and the accumulation and review of data against agreed acceptance criteria.

The level of validation should reflect the complexity of the process. The critical process parameters should be defined during the course of pre-formulation, pharmaceutical development and scale-up studies, and the validation protocol should challenge and explore them.

Prospective, concurrent or retrospective validation may be applied. Re-validation should be performed as identified per schedule and product .In some cases process validation may be conducted concurrently with performance qualification, for example, where an item of equipment is dedicated to one process producing one product.

Procedures
1. Analytical method
Analytical results should be accurate and reproducible.Critical factors that should be validated include:

- Specificity;
- Accuracy;
- Precision;
- Recovery;
- Linearity;
- System suitability for chromatographic determination and
- Robustness.

B. Storage Component
Storage material should be evaluated and selected for media, plasmids and reagents to provide the required properties of compatibility, stability, security and sterility.

C. Cleaning Validation
The establishment of acceptance criteria for contaminant levels in the sample should be practical and achievable. However, it is often considered that a cleaning procedure, that consistently reduces the contaminants to a level not exceeding one-thousandth of its lowest daily therapeutic dose in the highest daily therapeutic dose of the product, can be regarded as validated.

2. Computer Systems Validation.
A written validation plan should be available. Specifications should be identified and design review should be performed. The system should be tested (IQ, OQ and PQ should be performed and documented).

A. General
Aspects of computerized operations that should be considered include:
- Networks;
- Manual back-ups;
- Input/output checks;
- Process documentation;
- Monitoring;
- Alarms; and
- Shutdown recovery.

B. System Specification

The control document should contain the objectives of a proposed computer system, the data to be entered and stored, the flow of data, how it interacts with other systems and procedures, the information to be produced, the limits of any variable and the operating programme and test programme.

System elements in computer validation that need to be considered include hardware (equipment), software (procedures) and people (users).

C. Functional specification

A functional or performance specification should provide instructions for testing, operating, and maintaining the system, as well as names of the person(s) responsible for its development and operation.

The following general aspects should be kept in mind when using computer systems: location, power supply, temperature, and magnetic disturbances. Fluctuations in the electrical supply can influence computer systems and power supply failure can result in loss of memory.

D. Security

- Data should only be entered or amended by persons authorized to do so. Suitable security systems should be in place to prevent unauthorized entry or manipulation of data.
- The security procedures should be in writing. Security should also extend to devices used to store programmes, such as tapes, disks and magnetic strip cards. Access should be controlled.
- Traceability is of particular importance and it should be able to identify the persons who made entries/changes, released material, or performed other critical steps in manufacture or control.
- The entry of critical data into a computer by an authorized person requires an independent verification and release of use by a second authorized person.

E. Back-ups

Regular back-ups of all files and data should be made and stored in a secure location to prevent intentional or accidental damage.

F. Validation of hardware and software:

Hardware

The validation/qualification of the hardware should prove:

- The capacity of the hardware matches its assigned function (e.g. foreign language);
- That it operates within the operational limits (e.g. memory, connector ports, input ports);

- That it performs under worst case conditions (e.g. long hours); and
- Reproducibility/consistency (e.g. at least three runs covering different conditions).

Software

Software is the term used to describe the total set of programmes used by a computer which should be listed in the menu or main menu.

Records are considered as software with focus placed on accuracy, security, access, retention of records, review, double checks, documentation and reproduction accuracy.

HARDWARE	SOFTWARE

1. Types
   o   Input and Output device
   o   Signal converter
   o   Central Processing Unit (CPU)
   o   Distribution system
   o   Peripheral devices
	1. Level
	   o Machine language
	   o Assembly language
	   o High level language
	   o Application language

2. Key aspects
   o   Location
      environment
      distance
      input devices
   o   Signal conversion
   o   I/O operation
   o   Command overrides
   o   Maintenance
	2. Software Identification
	   o Language
	   o Name and Function
	   o Input and Output
	   o Fixed set point
	   o Variable set point
	   o Edits
	   o Input manipulation
	   o Programme overrides

3. Validation
   o   Function
   o   Limits
   o   Worst case
   o   Reproducibility/consistency
   o   Documentation and Re-validation
	3. Key aspects
	   o Software development
	   o Software security

	4. Validation
	   o Function
	   o Worst case
	   o Repeats
	   o Documentation
	   o Re-validation

Table 6: Summary of validation requirements for computer systems.

3. Cleaning Validation (General): The objective of cleaning validation is to prove that the equipment is consistently cleaned from product, detergent and microbial residues to an acceptable level, to prevent contamination and cross-contamination.

These guidelines address the general requirements on cleaning validation, excluding specialized cleaning or inactivation that may, e.g. be required for viral or mycoplasma removal in the biological manufacturing industry.

The validated procedure should be followed consistently, adhered to, appropriately documented and recorded in cleaning logs and maintained to ensure that the equipment is always cleaned as required.

Cleaning procedure should complies with the acceptance criteria, microbiological aspects (bio-burden control) and re-validation requirements.
Pharmaceutical products can be contaminated by a variety of substances such as disinfectants, and decomposition residues which include:
- o product residue breakdown occasioned by, e.g. use of strong acids and alkalis during the cleaning process; and
- o breakdown products of the detergents, acids and alkalis that may be part of the cleaning process.

The laboratory should have a strategy on cleaning validation:

- o product-contact surfaces;
- o cleaning after product changeover (when one pharmaceutical formulation is being changed for another, completely different formulation);
- o between batches in campaigns (when the same formula is being manufactured over a period of time, and on different days). It seems acceptable that a campaign can last a working week, but anything longer becomes difficult to control and define;
- o bracketing products for cleaning validation. This often arises where there are products containing substances with similar properties (such as solubility) or the same substance in different strengths. An acceptable strategy is to manufacture the more dilute form (not necessarily the lowest dose) and then the most concentrated form. There are sometimes "families" of products which differ slightly as to actives or excipients; and
- o periodic evaluation and re-validation of the number of batches required should be included.

At least three consecutive applications of the cleaning procedure should be performed and shown to be successful in order to prove that the method is validated.

The cleaning validation protocol should be formally approved by the Quality Unit and other appropriate management.

Records of the cleaning validation, which include all raw data of the test results together with, e.g. the cleaning record (signed by the operator, checked by production and reviewed by QA), should be kept and a final validation report should be prepared. The final outcome should be stated, e.g. "all the acceptance criteria were met".

Personnel/operators who perform cleaning routinely should be trained and should have effective supervision.

Equipment: Normally only cleaning procedures for product-contact surfaces of the equipment need to be validated. Critical areas should be identified, particularly in large systems employing semi-automatic or fully automatic clean-in-place systems.

Disposable equipment should be used for products which are difficult to clean, equipment which is difficult to clean, or for products with a high safety risk where it is not possible to achieve the required cleaning acceptance limits via a validated cleaning procedure.

Sampling: There are two methods of sampling that are considered to be acceptable. A combination of the two methods is generally the most desirable.

(a) Surface sampling: This direct method of sampling is the most commonly used and involves taking an inert material (usually cotton wool or similar) on the end of a probe and rubbing it methodically across a surface.
Factors that should be considered include the supplier of the swab, area swabbed, number of swabs used, wet or dry swabs, swab handling and swabbing technique.

The swab location is important, taking into consideration the material of the equipment (e.g. glass, steel) and the location (e.g. blades, tank walls, fittings). Advantages of direct sampling include:
- areas hardest to clean and which are reasonably accessible can be evaluated (leading to establishing a level of contamination or residue per given surface area); and
- residues that are "dried out" or are insoluble can be sampled by physical removal.

Some disadvantages of using swabs include:
- inability to access some areas;
- presumes uniformity of contamination surface;
- must extrapolate sample area to whole surface; and
- reproducibility is suspect due to the human involvement and extraction efficiency.

(b) Rinse samples

This indirect method allows sampling of a large surface, of inaccessible areas or those that cannot be routinely disassembled and provides an overall picture.Rinse samples give sufficient evidence of cleaning .

(c)Other methods: Batch placebo method

A less frequent sampling method, due to its high cost. The method relies on the manufacture of a placebo batch and then checking it for carry-over of the previous product. It is an expensive and laborious process.

Scrubbing by hand: Manual cleaning methods are difficult to replicate.

Clean-In-Place (CIP) systems
Critical areas, i.e. those hardest to clean, should be identified, particularly in large systems that employ semi-automatic or fully automatic CIP systems.

CHAPTER 8   EXAMPLES OF SOME STANDARD OPERATING PROCEDURES (SOP)

Center for Integrative Physiology and Molecular Medicine

SOP No: 1   Revised No: Title: Pages:
Written/Edited/ Compiled by: Date:
Supersedes:
Approved By: Date :   Effective On:

    Automated, mechanical or electrical or other types of equipment including computer and computer-controlled equipments shall function satisfactorily. Such equipments should be calibrated properly and regularly with proper documentation. Maintenance of equipment is an extremely important. But sometimes it is neglected. Cost of maintenance including inspection, lubrication and adjustment of equipment is negligible when compared with the cost of emergency repairs, rebuilding or overhauling of equipment. Preventive maintenance is a program of scheduled inspections of equipment and instruments resulting in minor adjustments or repair for the purpose of delaying or avoiding major repair and emergency or premature replacement of spare parts.

Benefits of Breakdown Maintenance
- Better quality results
- Identification of components showing excessive wear
- Greater Safety
- Fewer interruptions in service
- Lower repair costs
- Less standby equipment required

Center for integrative Physiology and Molecular Medicine

SOP No: 2   Revised No: Title: Pages:
Written/Edited/ Compiled by: Date: Supersedes:
Approved By: Date :   Effective On

CARE OF BALANCE

Model of the Balance : Ohaus GT 4800, A It is an electronic precision balance designed to be versatile, accurate and easy to operate.
Installation environment:

1. Environment should be free from excessive air currents, corrosives, vibrations, temperature or excessive humidity.
2. Place the balance in proper place.
3. Adjust the leveling spirit at the center by adjusting the two front wheels.
4. Connect to the correct voltage line as indicated on the power cord specification.
5. Test for power supply.
6. Read the switch buttons carefully e.g., Mode, On/Off/Tare, print, Calibration, setup, as per described in the operation manual.
7. Read the error codes carefully.

Calibration :
All balances have been calibrated before shipment. But the calibration should be checked before start using. Calibration may be influenced

1. Due to variation in gravitational fields in different latitude of earth,
2. Handling during shipment and
3. Changes in the working location.
4. Zero the balance momentarily by pressing On/Tare button.
5. Place a known weight on the canter of the pan. Displayed weight should not differ from the known weight. If differs recalibration is necessary.
6. Remove all weights from the platform and Enter into calibration menu by pressing and holding ON/Tare button then release immediately.
7.   [[]   ]   g will be displayed

8. Momentarily press ON/Tare and   —   [will be displayed. Place the required
    calibration weight on the platform of the balance.

9. Momentarily press on/Tare and after sometime calibration weight be displayed

10. Now the balance is calibrated.

Setup Integration: In **A.L.** menu by repeated pressing of the following will be displayed

Display	Integration Description	Response time
AL 0	Minimum	Maximum
AL 1	Reduced	Increased
AL 2	Normal	Normal
AL3	Maximum	Reduced

Integration can be selected by pressing **On/Tare.**

Stability level: This means a number of displayed weights are within selectable range of each other. By pressing Mode repeatedly the following will be displayed.

Display	Stability level
0	Reduced
1	Normal
2	Increased

Stability level can be selected by pressing **On/Tare.**

Select a unit: Unit selection allows selecting units in the same way. At SEL menu units can be selected by pressing Mode repeatedly. The following will be displayed and units can be selected by pressing **On/Tare.**

Mode indicator	Weighing mode
g	Grams
dwt	Pennyweight
ct	Carats
oz	Ounce
ozt	Ounce troy
lb	pounds
t	teels
pc	Parts counting
Units 1	Custom Units1
Units 2	Custom Units2
Units 3	Custom Units3
UNDER, ACCEPT, OVER	Check weighing

Mode repeated **Setup menu:** Read all the menus carefully Push **ON/Tare** button and release

SETUP Mode appears. Then by pushing Mode operational can be changed as follows order.

Center for integrative Physiology and Molecular Medicine

SOP No: 3   Revised No: Title: Pages:
Written/Edited/ Compiled by: Date: Supersedes:
Approved By: Date :   Effective On

CARE OF pH METER

pH is defined as the negative logarithm of hydrogen ion concentration. In other words at a high concentration e.g., 1mol/l=100 the pH will be equal to 0 at a low H + concentration e.g., 10 –14 mol/l the pH is equal to 14. In order to measure pH value, a measuring electrode (pH electrode, glass electrode) and a reference electrode are needed. At the glass electrode a gel layer develops when it comes contact with a defined buffer solution (inner buffer). The H+ ions either diffuse out of gel layer, or into the gel layer, depending on the PH value of the measured solution. In the case of an alkaline solution the H+ ions diffuse out, whereby a negative charge is established on the outer side of the gel. Since the glass electrode has an internal buffer with a constant pH value, the potential at the inner surface of the membrane is also constant during the measurement. The total membrane potential is a result of their difference between the inner and outer charge.

The reference electrode consists of a reface element that is immersed in a defined electrolyte. This electrolyte must be in contact with the measured solution usually through a porous ceramic junction. The reference electrolyte and the reference element (e.g. silver/ silver chloride), Define the potential of the reference electrode. Here it is important that the reference electrolyte has a high ion concentration, which results in a low electrical resistance. Ideally no reaction between the reference electrolyte and the measuring solution should occur. A 3 Molar KCL solution has been fount to fulfill these conditions over a wide temperature range.

Nowadays. The most common used electrode is the combination electrode, in which the glass electrode is concentrically surrounded by the reference electrolyte.

The pH measurement of vaccines and intermediate products in DTL is performed using such a combined glass / reference electrode. Before measurement the pH meter should be calibrated against two standard buffer solutions of pH 7.00 and pH 4.00 respectively.

Materials and reagents:

pH instrument, radiometer, Combined Glass/reference electrode, Temperature sensor Calibration buffer

Procedure: Initial operation procedure for the combined electrode

  a. Carefully remove the tape sealing the filling hole and the porous pin respectively without damaging the porous pin. Check porous pin and glass electrode shaft for cracks. Refill with saturated KCl solution if necessary. Make sure that no air bubbles are present below the two reference electrodes.

Initial operation procedure for the measuring instrument

b. Switch on the water bath 30 minutes before the use and place calibration buffers, test samples and temperature sensor in it. Connect the power cord to line supply. Switch on the personal computer and login. Click program manager and double click the pH-icon. Enter ok, Click 'tools' followed by "calibration pH meter"

Calibration

c. Rinse the electrode with distilled water and wipe with tissue paper
d. Immerse the electrode in the pH buffer 7.00
e. Strike CAL key again 1 CAL TEMP press PRINT for sending the actual temperature of buffer 1n to the computer.
f. Press CAL key and change the pH value into 6.98 by striking SHIFT 6.980 followed by STORE
g. Wait until the signal is stable and press PRINT the pH will be sent to the computer
h. Press CAL and the electrode's zero pH will be displayed for a moment after which 2 CAL TEMP is displayed
i. Rinse the electrode and immerse it in the pH buffer 4
j. Strike PRINT and the temperature will be sent to the computer
k. Press CAL and enter the accurate pH at 25 0C being 4.01 by striking SHIFT 4.01 STORE wait for a stable signal
l. When CAL key has been pushed, the electrode sensitivity will be displayed for a moment after which pH meter switches to the next calibration step
m. The system is now ready for direct pH measurement.

pH measurement of the sample

n. Rinse the electrode with distilled water and immerse it in the sample for at least 30 seconds. PH and temperature will be displayed.
o. On the PC go to METEN and fill in all fields
p. When finished double click pH
q. Wait for a stable signal and press PRINT at the pH meter sending pH and temperature of the sample to the computer and press OK
r. Go to tools and file save to store data
s. Go to file and print in order to print a fill

Cleaning of the electrode

t. Rinse the electrode with distilled water and immerse in the enzymatic detergent. Rinse it again with distilled water and immerse the electrode in buffer solution between pH 4 and 8.

Requirements for test validity

The calibration results should met the criteria set:

1. The pH of BUF 1 should be + or – 0.01
2. The pH of BUF 2 should be + or – 0.01
3. Zero pH should be between 5.7-7.7

4. Sensitivity should be between 90%-100%

Table 8 :□pH requirements for the several vaccines

Vaccine	Final bulk	Final lot
Pertussis	6.7-7.3	
Tetanus	6.0-7.0	6.0-7.0
Trivalent Polio	6.8-7.2	6.8-7.2
D (P) T	6.8-7.4	6.8-7.4
ARV	7.2	7.4

Center for integrative Physiology and Molecular Medicine

SOP No: 4   Revised No: Title:
Pages:
Written/Edited/ Compiled by: Date:
Supersedes:
Approved By: Date :   Effective On

CARE OF SPECTROPHOTOMETER

Theoretical basis :
The variation of the colour of a system with change in concentration of some component forms the basis of what we term colorimetric analysis. The colour is due to the formation of colored compound by the addition on an appropriate reagent . The intensity of the colour may be compared with that obtained by treating the known amount of the substance in the same manner. The spectrophotometer is consists of two component , spectrometer or monochromator that produces colored light of different wavelength, and photometer that reads the intensity of light produced be monochromator. Light consists of radiation waves of different wavelength giving rise to different colors.

Table 9: Wavelengths of colors

Ultra-violet:	$\angle$ 400 nm	Violet	400-450 nm
Blue	450-500 nm	Green	500-570 nm
Yellow	570-590 nm	Orange	590-620 nm
Red	620-760 nm	Infra red 7	760 nm

INFRARED	VISIBLE	ULTRA-VIOLET

Beer and Lambert's law: When monochromatic light falls upon a homogenous solution a portion of incident light transmits, a portion of light is absorbed and a portion is reflected. Lambert states that when a monochromatic light passes away through a transparent medium the rate of decrease in intensity with the thickness of the medium is proportional to the intensity of the light. Light absorption and the light transmission for monochromatic light is a function of the thickness of the absorbing layer. Concentration of the colored substance has the definite effect on the light transmission or absorption. i.e. the intensity of the beam of a monochromatic light decreases exponentially as the concentration increases arithmetically. We can find an equation

$$O.D. = \varepsilon ct \quad \text{when}$$

$O.D$ = Optical Density of light absorbed

ε = Molecular extinction coefficient

t = thickness of the path

Location : Spectrophotometer can be installed normal desk or bench. Rear end should be 100 mm ventilation and electrical connection.

Connection to mains: Read carefully the installation instruction regarding voltage requirement and earthing. To connect the instrument to the main power line, ensure the voltage and fuse setting are correct for the local use, Plug the main cable into the power socket and confirm the earth contact.

Routine maintenance:
5. Keep the interior of the instrument dust free, Keep the sample compartment clean and wipe out any split chemicals. Moisture cannot be allowed to leak into the instrument
6. Switch off and disconnect the instrument for the power line
7. Wipe the outer surface with a lint free cloth soaked with detergent solution
8. Wipe the outer surface with a dampen cloth. Wipe the surface with a dry cloth,
    Isopropyl alcohol may used for final wiping

Replacement of tungsten halogen lamp:
8. Keep the spectrophotometer at disconnected and cool state.
9. Follow the instruction given in the operator's manual to Lamp access panel and cover.
10. Carefully remove the lamp assembly and pull the ceramic socket pin of the lamp holder.
11. Wear gloves to fit the Tungsten Halogen Lamp. Touch only the base of the lamp or the mounting plate.
12. Insert the pin of the lamp with a clear tissue. Secure the lamp assembly according to the operator's manual.
13. Reconnect lamp cover, and fit all the screw of the lamp access panel.
14. Use the spectrophotometer after calibration properly.

Replacement of deuterium lamp:
8. Keep the spectrophotometer at disconnected and cool state.
9. Follow the instruction given in the operator's manual to Lamp access panel and cover.
10. Carefully deuterium Lamp from the supply socket and pull the ceramic socket pin of the lamp holder.

11. Wear gloves to fit the Lamp. Touch only the base of the lamp or the mounting plate.
12. Release the lamp assembly following operators manual
13. Insert the pin of the new lamp with a clear tissue. Secure the lamp assembly according to the operator's manual. Reconnect lamp cover, and fit all the screw of the lamp access panel. Use the spectrophotometer after calibration properly.
14. The UV radiation from a deuterium lamp is harmful to the skin and eyes. Wear apron and eye protectors.

Wavelength accuracy and repeatability:

10. Set power on and allow the warm the spectrophotometer and set wavelength at 351 nm at abs mode under closed lid.
11. Press Zero, insert the Holmium filter in the cell holder and close the lid.
12. Read and record the absorbance
13. Remove the holmium filter and select the wavelength mode
14. Increase the wavelength by 1 nm and record the wavelengths 351 to 371 nm.
15. Hopefully the peak reading will be at 361 $\pm$ 1 nm
16. Repeat the same test with Didymium filter within the wavelength range 527-to 547 nm .The peak will be at 537 $\pm$ 2 nm.
17. Repeat the same test with Didymium filter within the wavelength range 797 to 817 nm .The peak will be at 807 $\pm$ 2 nm
18. Repeat the test thrice. The allowable variation of the peak must not be greater than 1 nm.

Stray Light:

6. Set power on and allow the warm the spectrophotometer and set wavelength at 340 nm at abs mode under closed lid.
7. Lift the lid, insert the Blanking plate in the cell holder to block the beam and close the lid and set the dark current pressing zero.
8. Remove the Blanking plate and close the lid and press Zero.
9. Insert a 1A gauge attenuator into the cell holder and close the lid. And record the reading (A) and press zero, remove the attenuator.
10. Insert 50g/l the aqueous (Na NO$_2$) solution in the cell holder and record reading (B)
11. (A) + (B) should be greater than 3.0 A or
12. Insert 10g/l the aqueous (NaI) solution in the cell holder and record reading (B) at 220 nm.
13. Again Attenuator reading at 220 nm (A) + (B) should be greater than 3.0 A

Photometric accuracy:
1. Set power on and allow the warm the spectrophotometer and set wavelength at 546 nm at abs mode under closed lid and press enter.
2. Lift the lid, insert the Blanking plate in the cell holder to block the beam and close the lid and set the dark current pressing zero.
3. Remove the Blanking plate and close the lid and press Zero to set dark current.
4. Insert a zero absorbance neutral density filter into the cell holder and close the lid and record the reading. And then press zero.
5. Insert 2A calibration filter standard and compare the display reading with the previous reading. Variation should not be higher than $\pm$ 0.5 %, 02 A within 10 seconds in any case.

Setting an wavelength:
1. Set power on and allow the warm the spectrophotometer and wavelength 0f 340nm appears in the monitor. Let us adjust the wavelength at 695 nm
2. Touch the edit mode and first figure **(0)** of **0340** starts blinking, press enter, the second figure **(3)** of 0340 will blink, edit this figure 3 to desired number (say **6**), adjust 6 by pressing up/down and press enter..
3. Then the third figure **(4)** of 0340 will blink, edit the figure 4 to desired number (say **9**), adjust 9 by pressing up/down and press enter.
4. Then the fourth figure **(0)** of 0340 will blink, edit the figure 0 to desired number (say **5**), adjust 9 by pressing up/down and press enter.
5. By this way wavelength is set at **695** nm.

Reading of the sample:
1. By pressing up/down set the abs mode and take the cuvette having blank inside the cuvette holder,
2. Close the lid and press Zero to set Zero
    1. Take the cuvette having Test/ Standard sample inside the cuvette holder, close the lid and record absorbance
    2. Calculate the concentration of the test sample by comparing absorbance with that of standard.

Calculation :

$$C = \frac{\text{Conc. of the Standard} \quad x \quad \text{abs of the Test} \, x \quad \text{Dilution factor}}{\text{Abs of the Standard}}$$

Center for integrative Physiology and Molecular Medicine

SOP No: 5   Revised No: Title: CARE OF AUTOCLAVE   Pages:
Written/Edited/ Compiled by: Date:
Supersedes:
Approved By: Date :   Effective On

    Autoclave is an integral part of any Biological laboratory. Sterilization done by using an Autoclave. Good Sterilization Practice ensures that cultures, containers, media and equipment are treated in such a way that only desired organisms that are inoculated will grow and all others will be eliminated. These are done by the use of heat, chemicals, radiation and filtration. Heat proved to be most popular method for sterilization. There are two methods of thermal sterilization: moist heat (saturated steam) and dry heat (hot air).

    Relatively few chemicals are capable of performing sterilization and have the additional properties of stability, safety, lack of colour etc. Gaseous Ethylene oxide and formaldehyde, liquids such as glutaraldehyde and Hydrogen peroxide are used as chemical sterilizer.
    The wide application of sterilization process makes it mandatory to impose strict controlmeasures to validate the results obtained.

Process	Physical Method	Chemical Method	Biological Test Organism
Dry Heat	Temperature Recording Charts	Colour Change Indicator	*B. Subtilis var niger*
Moist Heat	Temperature Recording Charts	Colour Change Indicator	*B. searothermophilus*
Chemical Ethyleneoxide Formaldehyde Glutaraldehyde H_2O_2			Strips of *B. Subtilis var niger* *B. searothermophilus* *Clostridium soprogenes*

Table : Methods of Validation Process of Sterilization

Varius steps that can be taken to ensure its proper functioning.

a) 121°C for 15 mins and 15 lb/inch^2 cycle is generally used. 126°C for 15 mins and 15 lb/inch^2 cycle

8 REFERENCES

1. Biological standardization and control WHO/BLG /97.1
2. WHO Technical Report Series 800, 1990
3. United States Pharmacopoeia
4. British Pharmacopoeia
5. WHO Technical report series 822, 1992
6. Guide for a quality systems Manual in a control laboratory, WHO /VSQ /98.04
7. WHO policy statement WHO / V &B/ 00.09
8. Regulation of Vaccines: building on existing drug regulatory authorities WHO / V&B /99.10
9. Good Manufacturing Practices for Pharmaceutical products WHO TRS 823
10. Cell culture techniques ASM news 51(4): 170-1830
11. Journal of American Pharmaceutical Association 46, 299-301 (1957)
12. Animal facilities: The Laboratory Environment
13. GMP and the work force: GMP Committee, Japan 1991
14. Biosafety in Microbiological and Biomedical Laboratories, CDC, USA 1993
15. A Textbook of Quantitative Inorganic Analysis, 1968.
16. Biosafety in Microbiological and Biomedical Laboratories, CDC.NIH, 1993

ABOUT THE AUTHOR

Mohammad Kabir Ahmed received his PhD from the University of Tokyo and is currently serving as an Associate Professor in the Preclinical Department of Biochemistry, Faculty of Medicine , University of Kuala Lumpur Malaysia, Royal College of Medicine Perak. He headed the Drug Testing Laboratory, Served as National Consultant in UNIDO and FAO of the United Nations. He prepared eight national laboratories for 3rd party accreditation on scopes related to Microbial contamination and chemical residues in food and feed.

Dr. Shaker Uddin Ahmed received his MD from from BSSMU, Dhaka and is currently serving as an Senior Lecturer, Faculty of Medicine , University of Kuala Lumpur Malaysia.